**Innovation,
Competition, and
Government
Policy in the
Semiconductor
Industry**

Charles River Associates

Charles River Associates (CRA), located in Boston, Massachusetts, has been providing research and consulting support to decision makers in industry and government since 1965. CRA's work covers a wide spectrum, including the following areas: antitrust policy; commodity-market forecasting; combined economic/engineering feasibility for new ventures; communications; consumer behavior; economic development; fuel industries, electric power, and energy economics; industry regulation; international trade; minerals, metals, and durable-goods industries; regional economics; science and technology policy; transportation planning; urban and intercity transportation economics.

Innovation, Competition, and Government Policy in the Semiconductor Industry

Robert W. Wilson
Peter K. Ashton
Thomas P. Egan

A Charles River Associates Research Study

LexingtonBooks
D.C. Heath and Company
Lexington, Massachusetts
Toronto

Library of Congress Cataloging in Publication Data

Wilson, Robert W.
   Innovation, competition, and government policy in the semiconductor industry.

   "Charles River Associates research study."
   Bibliography: p.
   1. Semiconductor industry—United States. 2. Technological innovations—United States. 3. Industry and state—United States. I. Ashton, Peter K., joint author. II. Egan, Thomas P., joint author. III. Charles River Associates. IV. Title.
HD9696.S43U59   338.4'762138152'0973   80-8317
ISBN 0-669-03995-0

Copyright © 1980 by D.C. Heath and Company

All rights reserved. No part of this publication may be reproduced or transmitted in any form or by any means, electronic or mechanical, including photocopy, recording, or any information storage or retrieval system, without permission in writing from the publisher.

Published simultaneously in Canada

Printed in the United States of America

International Standard Book Number: 0-669-03995-0

Library of Congress Catalog Card Number: 80-8317

Contents

	List of Figures	vii
	List of Tables	ix
	Preface and Acknowledgments	xii
Chapter 1	**An Overview of Government Policies Affecting the Semiconductor Industry**	1
	Identification of Relevant Policies	3
	Analytic Framework for Policy Analysis	5
Chapter 2	**The Environment and Technology of the Semiconductor Industry**	11
	Environment	11
	Technology	25
Chapter 3	**Innovative Behavior by Manufacturers of Integrated Circuits**	37
	Review of Framework	38
	The Innovative Record	38
	Characteristics of Innovative Firms: The Views of Industry Executives	43
	The Relative Importance of Semiconductor-Firm Characteristics for Innovative Behavior	46
	Statistical Testing of the Effect of Firm Characteristics on Innovation	69
	Conclusions and Policy Implications	72
Chapter 4	**Strategies and Competition Among Integrated-Circuit Firms**	77
	Key Elements of Strategy	77
	Case Histories of Strategies and Competition	82
	Conclusions	101
Chapter 5	**The Effect of Strategies on Firm and Industry Performance**	111
	Relation of Strategies to Firm Performance	111

	Industry Performance	118
	International Performance	131
	The Feedback of Performance on Organization and Strategy	135
	Summary and Policy Implications	137
Chapter 6	**Analysis of Government Policies in the Semiconductor Industry**	**141**
	Differential Impact of Policies among Strategic Groups	142
	Procurement Policy	145
	Government Funding of R&D	151
	Antitrust Policy	155
	Trade Policies	158
	Tax Policies	164
	Manpower Policies	170
	Policy Interdependencies	172
	Analysis of the VHSIC Program	174
	Summary and Conclusions	177
Appendix A	**Glossary of Integrated-Circuit Technology**	**179**
Appendix B	**Financial Incentives in the Semiconductor Industry: The Pot of Gold that Few Ever Reached**	**185**
	Bibliography	**199**
	Index	**207**
	About the Authors	**221**

List of Figures

2-1	Worldwide Semiconductor Sales	12
2-2	Entry into the Semiconductor Industry Compared with Growth in Semiconductor Production: 1951-1967	14
2-3	Entry into the Semiconductor Industry Compared with Growth in Semiconductor Production: 1964-1977	15
2-4	Family Tree of Semiconductor Technology and End Uses	20
2-5	Estimated Decreases in Computer-Memory Cost: 1973-1983	27
2-6	Integrated-Circuit-Manufacturing Stages	30
4-1	Strategic Groups in Digital Integrated Circuits	108
5-1	The Demand Curve and Consumer Surplus	123
5-2	The Changes in Consumer Surplus that Correspond with Changes in Price	125
5-3	Estimated Demand Curve for MOS Dynamic RAMs	127
5-4	Effect of Shifts in True Demand Curves on Calculation of Consumer Surplus	128
B-1	Compensation of the Highest-Paid Executive: Semiconductor Industry versus Electrical and Electronics Industry (SIC-36) versus All Manufacturing Industries	187
B-2	Compensation of the Three Highest-Paid Executives: Semiconductor Industry versus Electrical and Electronics Industry (SIC-36)	188

List of Tables

1-1	Relevant Policy Questions	8
2-1	Distribution of U.S. Semiconductor Sales by End Use	19
2-2	Concentration of U.S. Domestic Semiconductor Shipments: 1957, 1965, and 1972	22
2-3	Leading U.S. Semiconductor Manufacturers: 1955-1975	23
2-4	Estimated World Sales of Semiconductors by Major International Semiconductor Firms: 1972, 1974, and 1976	24
3-1	Semiconductor Device Structures	40
3-2	Semiconductor Processes	41
3-3	Key Digital Integrated-Circuit Product Families: 1961-1975	41
3-4	Semiconductor Device Patent Applications: 1969-1976	42
3-5	Semiconductor Process and Packaging Patent Applications: 1969-1976	43
3-6	Semiconductor Systems and Applications Patent Applications: 1969-1976	44
3-7	Distinguishing Characteristics of the Most Innovative Semiconductor Firms	45
3-8	Proportion of Top Executives' Time Spent on Matters Related to Innovation and Number of Key Innovations	49
3-9	Proportion of Top Executives' Time Spent on Matters Related to Innovation and Degree of Risk Taking	50

3-10	Comparative Labor Intensity: U.S. Domestic Semiconductor Manufacturing versus All U.S. Manufacturing: 1967, 1972-1976	55
3-11	Comparison of Cumulative R&D Expenditures, Semiconductor Device Patent Applications, and Significant Innovations for Nine Firms: 1972-1976	57
3-12	Responses to Questions Concerning Changes in Risk Over Time	58
3-13	Level of Risk Taking, Number of Key Innovations, and Number of Patents	60
3-14	Statistical Effect of Firm Characteristics on Key Innovations and Patents	72
5-1	Definitions of Performance Variables	113
5-2	Regression Results for Share and Growth Rate of MOS Integrated-Circuit Sales	115
5-3	Regression Results for Share and Growth Rate of Bipolar Integrated-Circuit Sales	115
5-4	Regression Results for Share and Growth Rate of Digital Integrated-Circuit Sales and Corporate Profitability	116
5-5	Test Statistics for the Hypotheses that Major-Innovation and Full-Line Product Strategies Yield Greater Performance	118
5-6	Net Earnings after Taxes as a Percent of Sales: 1964-1977	120
5-7	Net Earnings after Taxes as a Percent of Equity: 1973-1977	121
5-8	Comparison of Consumer Surplus and Sales for MOS Dynamic RAMs	130
5-9	Estimated Share of U.S.-Based Companies in Total World Shipments of Discrete and Integrated-Circuit Semiconductors: 1974-1977	131
6-1	U.S. Shipments of Semiconductors for Defense Consumption: 1963-1977	146

List of Tables xi

6-2	U.S. Government Semiconductor R&D Expenditures: 1955-1961	154
6-3	World Apparent Consumption of Semiconductors	159
6-4	U.S. International Trade of Semiconductors: 1967-1977	160
6-5	Adjusted U.S. Trade Data for Semiconductors: 1970-1975	162
B-1	Total Direct Compensation of Senior Officers at Selected Semiconductor Firms: Fiscal Year 1978	186
B-2	Average Annual White-Collar Salary in Selected Industries	189

Preface and Acknowledgments

The genesis for this book was concern among policy makers that the rate of innovation in technology-based industries has declined, thereby contributing to inflation, slow economic growth, and a deterioration in competitive performance by U.S. firms in international trade. As part of a diffuse response by the federal government to the policy issues raised by these developments, in September 1978 the Experimental Technology Incentives Program (ETIP) contracted with Charles River Associates (CRA) to study the innovative behavior and performance of firms in technology-based industries.

Among the objectives of the CRA study were (1) developing a methodology that policy makers can use to gain an increased understanding of the relationships between federal policies and the behavior and performance of firms in technology-based industries; (2) testing and refining this methodology in case studies of several such industries; and (3) identifying for further research and policy experiments important policy issues and problems in these industries. A companion volume, *Innovation, Competition, and Government Policy: A Framework for Analysis* (CRA, 1980), develops an analytical framework and methodology to assist policy makers in identifying and assessing the effects of federal policy in technology-based industries. ETIP and CRA selected the semiconductor industry, and digital integrated circuits in particular, as the first case study in this research because of the high rate of technological progress in that industry and heightened policy interest centering on the industry's international competitive position and its critical relationship to the rest of the economy.

Although various books, government reports, and the trade press were important in developing our understanding of the semiconductor industry, our most valuable sources of information and insights about the industry were personal interviews. We conducted structured interviews with executives of nine semiconductor firms, officials of the Semiconductor Industry Association, and members of venture-capital firms that have invested in the semiconductor industry. We also talked informally with several consultants to the semiconductor industry and with a few former employees of semiconductor firms.

We would like to thank the large number of people who graciously made themselves available for interviews and comments during this study. Executives of integrated-circuit manufacturers included W.J. Sanders III of Advanced Micro Devices; C. Lester Hogan of Fairchild; Edgar Sack of General Instrument; Robert Noyce of Intel; William Howard of Motorola;

Floyd Kvamme, John Nesheim, and Joe Van Poppelen of National Semiconductor; William Hittinger and Howard Rosenthal of RCA; Klaus Volkholz of Signetics; and Jim Comfort, Charles Phipps, and Dean Toombs of Texas Instruments. Executives of venture-capital companies included Rich Barry of Candela Electronics; Dennis Van Ness of Hambrecht and Quist; Eugene Kleiner of Kleiner, Perkins, Caufield and Byers; Richard Kramlich of New Enterprise Associates; and Robert Perring of Wells Fargo Investment Company.

Tom Hinkelman of the Semiconductor Industry Association provided very helpful advice and comments at an early stage of the study. Our discussions with Phil Ferguson of Ferguson Associates and Fred Zieber of Dataquest were also extremely useful. We would also like to thank Dataquest, Inc., for making material from their Semiconductor Industry Service available to us, and Integrated Circuit Engineering Corporation, for making copies of their publication *Status* available.

Members of the ETIP review panel also provided many helpful comments. In particular, Jack McMullen, Colin Mick, Bob Scace, Ted Schlie, and Greg Tassey went to extra effort to review material and provide assistance. Murray Bullis of the National Bureau of Standards also provided helpful guidance at an early stage of the study. Finally, Zvi Griliches of Harvard University and Merton Peck of Yale University have made valuable contributions throughout the study, and Richard Levin of Yale made very helpful comments on the draft final report.

A number of people at CRA made important contributions to the study. Bob Larner supervised the study, participated in the interviews and research, and made many helpful editorial revisions. In many respects his contribution was equal to that of the authors. David Howe, Marvin Lieberman, and Tom Reindel contributed material and research support for chapters 2 through 5. Their contributions were essential to the quality and timely completion of the study. Louis Caouette and Leslie Meyer made important contributions early in the study to chapters 2 and 3 and appendix B, and they both aided the interview and survey efforts considerably. Lynne Graybeal provided research assistance for the study, principally for chapter 6, and drafted some of the material in that chapter. Several other people also contributed to the early stages of the study. Jim Dalton helped in organizing the early research after having authored the major portions of the companion framework volume. Barbara Bahlke and Jonathon Levy undertook research into the industry trade literature, and the information they assembled was an important source for the final stages of the study. Janice DiNatale served as project secretary throughout the study and was instrumental in organizing the project files and library, handling communications, and typing draft material. Mary Ann Buescher provided excellent editorial supervision over numerous drafts of the manuscript. Other

Preface and Acknowledgments

members of the CRA Publications Department who made valuable contributions to the production of the manuscript include Kathy Davenport, Janet Fearon, Mary Margaret Franclemont, Jean Fried, Nicole Harris, Ellen Knox, Sharon Nathan, Susan Parker, and Bob Scheier.

While the contributions of all these people improved our study, none of them is responsible for any shortcomings. Except where specifically noted otherwise by a quotation or reference, the views contained in this volume belong only to the authors and do not necessarily reflect the views of any of the individuals or organizations cited here. The division of labor among the authors was roughly as follows: Robert Wilson had primary responsibility for chapters 3 and 5, wrote portions of chapter 4, and wrote the final version of the environment section of chapter 2; Peter Ashton had primary responsibility for chapters 1 and 6; and Tom Egan wrote the technology section of chapter 2 and portions of chapter 4 and also contributed substantial revisions to chapter 3.

1 An Overview of Government Policies Affecting the Semiconductor Industry

Within the past few years the semiconductor industry has been undergoing important changes. As circuit density has increased, semiconductor technology has become even more complex than before. Reflecting this increased complexity, the cost of an efficient wafer-fabrication facility has tripled in the last five years and is expected to double again by 1985. As entry of new firms and expansion of established firms have become more expensive, semiconductor manufacturers have outgrown the lending capability of the venture-capital market and have turned to other sources of capital. These new sources of capital are frequently either domestic end users or foreign manufacturers of semiconductor devices.

At the same time that foreign firms are investing in the U.S. semiconductor industry, import competition is becoming significant, especially in some of the more advanced commodity devices. In addition, the semiconductor industries of Japan, Germany, France, and the United Kingdom, with the financial support of their respective governments, have undertaken major research and development (R&D) efforts to develop advanced semiconductor technology and capture a larger share of worldwide sales from U.S. firms. This threat from foreign competition in an industry historically dominated by U.S. firms, combined with a concern over whether U.S. semiconductor firms and executives continue to have the entrepreneurial incentives and access to financial resources that drove the industry in its first three decades, has led both industry spokespersons and federal policy makers to reassess the desirability of a laissez faire posture toward the industry.

The federal government has played a significant role in the development and growth of the semiconductor industry. Two important characteristics of federal policies affecting the semiconductor industry have been their changing role and importance in different stages of the industry's development over time and their differential impact on different types of semiconductor firms. This chapter identifies those government policies that have had important effects on the industry and sketches the trends in policy over the thirty-year history of the industry. In chapter 6 we return to these policies and apply a framework for policy analysis to assess policy impacts on the semiconductor industry and its member firms.

The changing role and importance of policies over time can be seen by dividing the semiconductor industry into three distinct periods. During the early, youthful years of the industry (the 1950s and early 1960s), extensive interaction existed between government agencies (particularly the military and NASA) and firms in the industry. This interaction arose from the government's great interest in military and space applications of semiconductor technology and its willingness to fund company R&D directed toward improved, more reliable devices. In addition to this push to semiconductor technology, the government also exerted a demand-pull influence. The federal government was the industry's principal customer, and the profits offered by procurement contracts attracted many new firms to the industry.

The second distinct period, beginning in the early 1960s and continuing into the late 1960s, may be described as one of laissez faire. During this time government demand, while still important, decreased rapidly in relative size compared to the total demand for semiconductors. Although the military and NASA remained interested in the development of new devices, new commercial applications were found that were more attractive to many of the firms in the industry. As a result, the "pull" from the government demand decreased in overall importance. At the same time, no new policies were adopted or implemented that had negative influences on firms' incentives to innovate (Webbink 1977).

The third distinct period, which started in the late 1960s, can be characterized as one of increased frustration by industry executives with government policy. Government tax and trade policies have played an increasingly important role in the semiconductor industry. As the industry grew worldwide in scope, foreign competition heightened, spurred in part by the support that foreign governments have given to their semiconductor industries. Various changes in policy (or lack of policy responsiveness in some instances) have led U.S. semiconductor executives to view the role of government policy with a critical eye, and the once cordial relationship between government and the semiconductor industry has now taken on a more adversarial tone (see Hogan 1979).

In addition to the changing nature and importance of policy over time, there is also evidence that policies have had different effects on different semiconductor firms, depending on their resource base, product mix, innovative behavior, and competitive strategies. This differential impact of policies, which has been the case not only in recent years, but also in the early years of the industry, has important implications for policy makers. Because policies may favor or impact more directly on some firms in the industry than others, policy makers must be cognizant of the effects a policy can have on industry structure and the balance among different types of firms within the industry.

An Overview of Government Policies

Identification of Relevant Policies

Six major government policies or groups of policies have had a substantial impact on the semiconductor industry. The list of relevant policies was determined based on our analysis of the industry's behavior, the economics and trade literature on the industry, and interviews with industry executives and government officials. These six sets of policies are as follows:

Procurement

R&D funding

Antitrust

Trade
   export policies
   import policies

Tax

Manpower

All six policy types are of a general nature and affect industries throughout the economy, although three—procurement, government funding of R&D, and antitrust—had specific applications in the semiconductor industry. Nevertheless, the impacts of these policies have differed across industries and among firms within the same industry. These policies are briefly summarized in the next paragraphs. In chapter 6 we provide a more detailed analysis of their impacts on the semiconductor industry and illustrate how our framework for policy analysis can be used to assess the effects of government policy in an industry.

Procurement. Procurement policy is the acquisition by the government from private firms of goods and services for its own use. The Department of Defense and NASA have traditionally accounted for the bulk of the government's procurement of semiconductors. Military use of semiconductors was extremely important in the 1950s and the early 1960s, particularly in the years following the invention of the integrated circuit.

R&D Funding. Government funding of industry R&D also aided the early growth and development of the industry. Government R&D funding have usually taken the form of a contract (similar to procurement), in which the government agrees to fund research conducted by private firms in return for the government's right to all research results. R&D may also be performed *by* the government, and the results are provided to firms for their general use (given that the work is not kept secret for security reasons). Government-performed R&D was fairly prevalent in the 1950s. For example, the National Bureau of Standards (NBS) conducted detailed studies of miniaturization in Project Tinkertoy, and both the Army and Air Force also did research. However, on the whole, government research was not nearly as significant as research conducted by the industry (Kleiman 1977).

Antitrust. Antitrust policy has been an important factor in the semiconductor industry in two distinct periods. The first was during the youthful years of the industry when the Justice Department brought suit against AT&T and in 1956 entered into a consent decree, which prohibited Bell Laboratories and Western Electric from selling semiconductors in the merchant market. Other large manufacturers of receiving tubes, which had entered the semiconductor industry, may also have been deterred from expanding as rapidly as possible into this new field for fear of raising antitrust problems (Tilton 1971). Antitrust policy has recently come under discussion again, with U.S. firms calling for the relaxation of antitrust restrictions to enable them to compete more effectively with the large European and Japanese conglomerates that manufacture semiconductors.

Trade. Trade policies have also taken on increased importance in the industry as foreign demand for semiconductors has increased and foreign firms have also begun competing in the United States. Trade policies refer to both import and export policies controlling the flow of intermediate and finished products into and out of the United States. Foreign competition has increased markedly during the 1970s, with many of these foreign firms receiving significant financial support from their governments. Imports of semiconductors into the United States increased 45 percent a year between 1970 and 1975, while exports grew at an average annual rate of only 17 percent when the figures are adjusted to subtract the value of devices under U.S. tariff schedules 806.30 and 807.00. (U.S. DOC 1979). This rapid rise in imports relative to exports results in part from difficulties U.S. firms have encountered in selling in foreign countries (for example, Japan), in contrast to the experience of foreign firms that have had little problem selling in the United States. Part of the rapid rise in imports, however, can be attributed simply to imports starting in 1970 from a smaller base than exports.

Tax. Tax policies include a whole range of different policies enacted for various reasons. Some of these policies have had important impacts on the semiconductor industry, particularly in recent years. Tax policies may be specifically designed to stimulate investment in R&D or innovation, or to be more general, dealing with the way firms allocate and treat their expenditures. General tax policies affecting the semiconductor industry have included the Investment Tax Credit (ITC), tax treatment of stock options, expensing versus capitalizing R&D expenditures, treatment of foreign-earned income, and (perhaps most important) the capital-gains tax. Recent changes in many of the laws dealing with these taxes have had a strong effect on the semiconductor industry.

Manpower. Government policies regarding manpower and labor training have been important to the semiconductor industry since skilled labor is a particularly scarce resource in the industry. Government support of universities has increased the supply of trained engineers to the industry, in addition to stimulating considerable research at the university level. The supply of qualified manpower, in fact, tends to fluctuate with the amount of government research funding allocated to universities. Since the industry's demand for skilled engineers has been steadily increasing, these fluctuations in supply have sometimes resulted in shortages (*Electronics News* 17 December 1979).

Finally, it should be noted that other government policies that affect industry generally have not had strong impacts in the semiconductor industry. In particular, patents and government regulatory policies have had little significant impact on the industry. One reason for the limited effects of patent policy is that rapid technological change in the industry has negated the monopoly or proprietary effect of many patents. Moreover, licensing of patents on a very liberal basis has precluded a firm or a group of firms from using their patent holdings to gain market power. In addition, federal regulatory policies such as EPA or OSHA rules and standards, which are alleged to have retarded growth and innovation in some industries, have until recently had little practical significance for the semiconductor industry.

Analytic Framework for Policy Analysis

The analytical framework for assessing the microeconomic effects of government policies is based on an analysis of the factors affecting a firm's decisions to innovate and the impact of these decisions on the firm's performance. For a more complete description of this framework, see chapters 2 and 6 of CRA (1980). Innovative behavior is considered part of a firm's overall competitive strategy, which in turn is conditioned by its corporate goals, the organization of the firm, and the environment in which the firm operates. A brief description of the elements of our framework and how they affect firm and industry performance is given in the following paragraphs. Application of the framework to the semiconductor industry and a detailed analysis of government policies with important effects on the industry are presented in chapter 6.

Corporate goals encompass specific objectives that serve as yardsticks by which management can evaluate the firm's performance. Examples of corporate goals are a target rate of return, target market shares, or a target growth rate in sales or profits.

The firm's organization includes both the resources of the firm and the way the firm structures its resources. Tangible resources of the firm include individuals with various kinds of education and experience and various levels of skills (including R&D skills), plant and equipment, production materials, and financial resources. Intangible resources include the firm's managerial philosophy, leadership style and quality, attitudes toward risk, and good will.

The firm's environment consists of the elements outside the firm that provide it with opportunities and constrain its behavior. These elements commonly include trade associations, the structure of customer markets, opportunities for innovation, legal and societal constraints, the state of the economy, and market structure. Market structure encompasses several dimensions including the number and size distributions of buyers and sellers, entry conditions, the degree of product differentiation, market growth, foreign competition, importance of fixed costs, and structure of input markets.

In formulating its strategies, a firm must assess its comparative strengths and match its capabilities to the perceived opportunities in the environment (Rosenbloom 1978). Innovative and competitive behavior are determined by these strategies. Innovative behavior involves changes in product features and production processes and includes both original innovation and imitation of innovations made by other firms. Competitive behavior refers to the interaction among rivals and involves the implementation of the firm's product, marketing, and pricing strategies within the context of rivalry.

Performance represents the final outcome of the competitive behavior of firms. Firm performance is the outcome realized by the firm as a consequence of the implementation of its strategies and of its interaction with rivals. Aspects of firm performance are profitability, growth, and change in market share. Market performance is the product of the interactions of all firms and includes efficient allocation of society's resources and the industry's rate of technological progress.

Policy represents the collection of all government programs and actions that affect the performance of firms. Government programs may affect industry directly or indirectly, and policies may be tailored to specific industries or be of a more general nature.

The framework for policy analysis is derived from this general schema of relationships and is developed through a series of questions for policy makers to ask about the impact of a particular policy on each element of the

An Overview of Government Policies

schema. These questions are presented in table 1-1. The proposed questions focus on the impact of a change in an element of the framework on the firm's incentives to innovate, the probability of innovation, and the form innovation will likely take. These questions will help policy makers to determine how policy affects variables that influence the decision to innovate, particularly the rewards and risks of innovating.

Government policies are part of a firm's environment and can affect performance through their effects on the various determinants of performance. In chapter 6 we analyze the effects of policy on performance by tracing the impact of policy on intermediate variables in the framework such as innovative or competitive behavior and the effects, in turn, of changes in these variables on performance.

Policies may influence a firm's goals either by creating new goals or reinforcing existing goals. Policies may also induce changes in the firm's organization: the structure and resource base of the firm may be altered either to encourage or discourage innovative behavior or to heighten or lower barriers to innovative activity.

Since firms may pursue different product, price, and marketing strategies, government policies will affect different firms differently. Firms may be grouped according to the similar strategies that they follow, and analyzing the impacts of policies on each of these different strategy groups is important. This type of analysis can identify the effects of policy on the innovative and competitive behavior of different firms within the industry and ultimately on firm and industry performance.

Once this initial analysis of policy impacts has been made, further analysis of the *interactions* among policies must be made. Policies often are interdependent — that is, they do not work their effects in isolation but rather interact with each other. Some policies either reinforce or offset the effects of other policies. These interdependencies can be analyzed by determining which elements of the framework each policy directly impacts and how these direct impacts may be interrelated. The net effects may be very different from those anticipated by policy makers, and thus it is important to understand and evaluate each policy in the context of its relationships with other policies affecting the firm.

In the following chapters, we apply this framework and methodology to a case study of the semiconductor industry. Chapter 2 describes the environment and technology of the semiconductor industry and contains background material that we draw on in later chapters. Chapter 3 deals with innovative behavior by manufacturers of digital integrated circuits. The reader should be alerted that the product focus of our report shifts from time to time. In some parts of the report, such as when we are discussing the broad effects of government policies, the analysis and data pertain to the semiconductor industry as a whole. In other parts of the report, such as

Table 1-1
Relevant Policy Questions

Elements of the Schema	Questions
Environment	
Demand factors	How does the policy affect demand for the aggregate output of the industry?
	Does the policy affect the output *mix* demanded from the industry? Does the policy stimulate the demand for certain of the industry's products relative to others?
	Does the policy affect the market risk faced by firms?
Supply factors	How does the policy affect the technology base of the industry?
	How does the policy affect the opportunities for innovation perceived by firms in the industry?
	Does the policy affect the technical risk faced by firms?
	Does the policy affect the supply of skilled resources to the industry?
	Does the policy alter the availability or cost of capital to firms or affect the expectation of investors or the functioning of the capital market?
	Does the policy induce changes in market structure?
	Does the policy affect the competitiveness of domestic firms relative to foreign firms?
	Does the policy affect the ability of firms to appropriate the returns from innovation?
	How does the policy affect the incentives to technical innovation of key entrepreneurs and technical personnel in the industry?
	How does the policy interact with other policies and regulations in the industry? Does this interaction reinforce or diminish the forces set in motion by the policy?
Corporate goals	Does the policy put pressure on firms to pursue new goals?
	Does the policy reinforce firms' existing goals?
Organization	Does the policy alter the prestige or relative position in the corporate hierarchy of the individual(s) responsible for decisions on R&D and innovation?
	Does the policy alter management attitudes toward risk?
	Does the policy allow the firm to alter the mix of skills in its resource base?
Strategy formulation (given an assumed mix of strategies)	Will existing mixes of strategies be reinforced or altered? What is the likely net effect on resources devoted to process and product innovation?
	How will rivals respond to changes in the mix of strategies by certain firms?

An Overview of Government Policies

Table 1-1 continued

Elements of the Schema	Questions
	How will the policy affect the number and size distribution of strategy groups, strategic distance, and market interdependence?
	If current strategies are associated with poor performance, how will the firm alter its mix?
	How does the policy affect the incentives for innovation at the firm level? (This aspect can be examined by projecting effects on cost and returns from investment in innovation.)
Performance	How does the policy affect social performance? Are changes in technology, prices, costs, quality, and diversity of product offerings the result of the policy?
	Are undesirable externalities ameliorated?
	How does the policy affect firm performance (for example, profits, growth)?
	Does the policy generate spillovers, desirable or undesirable, in other industries?

when we are discussing the innovative behavior or competitive strategies of semiconductor firms, our focus narrows to digital integrated circuits or to particular semiconductor devices [for example, bipolar logic, large-scale integrated (LSI) memories, or microprocessors]. The chapter draws heavily on the semiconductor trade literature and interviews that the CRA staff conducted with executives of semiconductor firms and other sources knowledgeable about the industry and assesses the relative importance for innovation of various firm characteristics identified in our framework. Chapter 4 examines the competitive strategies followed by different firms in the industry. The chapter is organized around the competitive races to develop and improve products in five different areas: bipolar logic, custom LSI circuits, LSI memories, microprocessors, and circuits for consumer products.

Chapter 5 describes and analyzes the financial and competitive performance of the semiconductor industry and its member firms. The strategies that contributed to the success or failure of firms in the five "product races" discussed in chapter 4 are identified and evaluated. Finally, chapter 6 utilizes the framework for policy analysis to examine the effects on the semiconductor industry of the policies we cited earlier.

2 The Environment and Technology of the Semiconductor Industry

The semiconductor industry stands out among modern industries as having experienced extremely rapid technological change and growth. As a background for the analysis in subsequent chapters, this chapter discusses the environment and technology of the semiconductor industry, and particularly the digital integrated-circuit portion of the industry. The first section discusses the environment of the semiconductor industry. Major topics are the life cycle, opportunities for innovation, growth, and entry into the industry; end uses of semiconductor devices and the sensitivity of the industry to business cycles; and some remaining aspects of the structure of the industry including concentration, foreign competition, and vertical integration. The second section examines the technological principles of the semiconductor industry that permit rapid price declines and foster innovative behavior. It discusses the technological concept of the transistor, the digital integrated circuit, the experience curve, and increasing circuit complexity.

Environment[a]

*The Life Cycle of the Industry:
Opportunities for Innovation, Growth, and Entry*

An important determinant of industry behavior and changes in industry structure over time is the life cycle of an industry. Over the long sweep of its history, an industry often passes through three stages of development. The first is the growth stage, when the industry's technology and products are new and undergo rapid change. Firms are penetrating new markets. Opportunities for innovation are relatively plentiful and the rate of growth is high. In the second stage the industry attains maturity. Sales are mainly for replacement rather than for new applications. Major product innovations are infrequent, capital investment increases, and competition focuses on cutting production costs. As the growth rate slows, marginal firms exit the industry, concentration increases, and the industry settles into a mature mode. A third stage may arise if this maturity is disrupted by new products

---

[a]Leslie Meyer and Tom Reindel wrote major portions of this section.

arising from technological advances outside the industry or by major exogenous declines in demand caused by changing tastes or demographic changes. In the declining stage there is a further shake-out of firms.

Viewed in terms of a life cycle, the semiconductor industry has been and continues to be in a stage of rapid growth. Figure 2-1 shows the rapid growth of the worldwide industry since 1959. During the first decade, sales

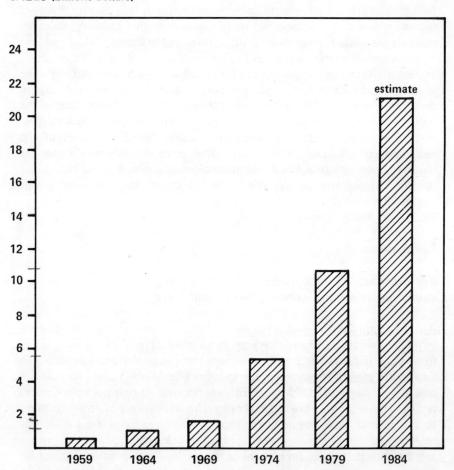

Sources: 1959, 1964, and 1969 data from Douglas W. Webbink, "The Semiconductor Industry: A Survey of Structure, Conduct, and Performance," Federal Trade Commission Staff Report, January 1977, p. 11.

1974, 1979, and 1984 data from *Business Week*, "Can Semiconductors Survive Big Business?" 3 December 1979, p. 68.

Figure 2-1. Worldwide Semiconductor Sales

of semiconductor devices worldwide more than quadrupled, from $395 million in 1959 to $1.7 billion in 1969. Semiconductor sales have almost doubled during the period from 1974 to 1979, from $5.4 billion to $10.5 billion. The predicted 1984 sales are $21.5 billion. Since unit prices have on average fallen at the rate of 10 percent per year, these dollar figures understate the growth rate in the volume of semiconductors.

The rapid growth of the semiconductor industry has been spurred by abundant opportunities for innovation, which have brought declining costs per electronic function and increasing capabilities of semiconductor devices. The transistor, the most fundamental building block of integrated-circuit technology, resulted from half a century of progress in solid-state physics that accelerated during World War II.

Bell Laboratories played an important role in the post-war semiconductor revolution. A large team of gifted scientists was assembled by the firm to work on solid-state physics, the objective being to produce devices useful to the telecommunications industry. Since germanium and silicon devices were used during World War II for microwave detection subsequent study focused on these conducting materials. On 23 December 1947, the first germanium transistor was developed after a great deal of collaborative effort by physicists, chemists, metallurgists, and engineers (Braun and MacDonald 1978).

In 1954 silicon replaced germanium in the production of transistors. Silicon was a more abundant material from which transistors could be made to work reliably at high operating temperatures. The use of silicon led to the introduction of the planar process, which allows transistors to be made in a batch-production process. In the late 1950s and the early 1960s the integrated circuit was developed and introduced. This development progressively allowed tens and hundreds of thousands of transistors and diodes to be placed upon a single chip of silicon about one-fourth of an inch on a side.

Technological change and growth have contributed to a high rate of entry in the semiconductor industry. Figures 2-2 and 2-3 show the number and pattern of entrants between 1951 and 1977 for a sample of 90 semiconductor firms for which entry dates could be obtained. The pattern of entry exhibited in figures 2-2 and 2-3 reveal roughly three periods of peak entry: 1952-1953, 1959-1963, and 1968-1972. In addition, the number of entrants since 1972 has declined severely.

Industry growth has certainly played an important role in the absolute number of entrants into this industry. Growth in applications and sales of semiconductors provides an environment conductive to new entrants. However, increases in the rate of growth of semiconductor shipments seem to be inversely correlated with the pattern of entry as exhibited in figures 2-2 and 2-3. More specifically, the peak periods in entry precede the significant periods of growth by approximately five years.

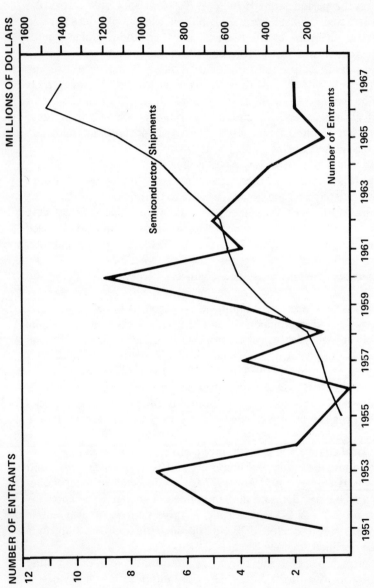

Sources: A sample of ninety semiconductor firms for which entry dates could be found were obtained from the following sources: Golding (1971, pp. 154-171, 242-244); Tilton (1971, pp. 52-53, 79); Braun and McDonald (1978, pp. 61-62, 67); Industrial Technology Hearing (30 October 1978, p. 91).

Semiconductor production figures for 1955 through 1959 are from Tilton (1971, p. 90). Figures include captive production. Data from 1960 to 1967 from U.S. Department of Commerce, 1979. *Report on the Semiconductor Industry*, Washington, D.C.: U.S. Government Printing Office, September, p. 39.

Figure 2-2. Entry into the Semiconductor Industry Compared with Growth in Semiconductor Production: 1951-1967

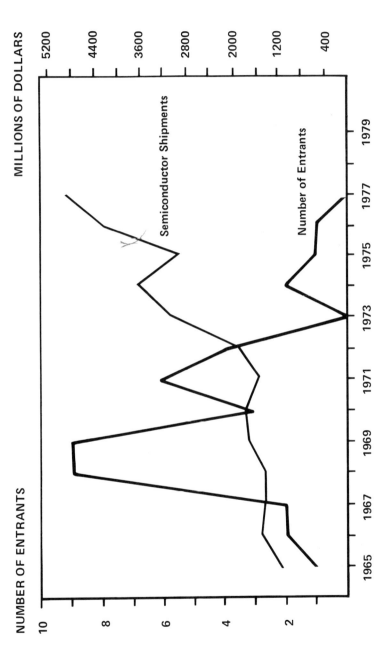

Figure 2-3. Entry into the Semiconductor Industry Compared with Growth in Semiconductor Production: 1964–1977

Sources: A sample of ninety semiconductor firms for which entry dates could be found were obtained from the following sources: Golding (1971, pp. 154-171, 242-244); Tilton (1971, pp. 52-53, 79); Braun and McDonald (1978, pp. 61-62, 67); Industrial Technology Hearing (October 30, 1978, p. 91).

Semiconductor Shipments from U.S. Department of Commerce, 1979. *Report on the Semiconductor Industry*, Washington, D.C.: U.S. Government Printing Office, September, p. 39.

These results are reasonable given that firms need three to five years from their foundation to gear up to full production. Firms are founded with the exepectation that the rate of growth of demand is going to increase significantly in the near future. In addition, these expectations might have a self-reinforcing feedback effect in that large numbers of new entrants will contribute to increases in demand and production.

The high rate of technological change, like the high growth rate experienced by this industry, has also contributed to the large absolute number of entrants. The degree to which new semiconductor innovations will provide potential entrants with an advantage depends to a large extent on the degree to which their new innovations obsolete established firms' products and production technology.

The histories of successful entrants illustrate the role new products or processes played in those entrants' successes. Still a young firm in 1954, Texas Instruments, (TI), building on research at Bell Laboratories, introduced the first silicon transistor. TI's lead in this development lasted for three years and catapulted it to the top of the industry by 1957. Transitron, the second largest transistor firm in the late 1950s, also had its growth and success built on a new product: the gold-bonded diode. Although the gold-bonded diode was first developed at Bell Laboratories, Transitron was ". . . the first to work out many problems associated with large-volume production and to achieve yields high enough to permit a price competitive with the less reliable point-contact diode then in use . . ." (Tilton 1971, pp. 66-67).

In the late 1950s, both Fairchild and Motorola entered the semiconductor industry and rose to success on ". . . the wave of new technology using silicon and relying on the oxide masking, diffusion, planar, and epitaxial techniques . . ." that began in the early 1960s (Tilton 1971, p. 65). While TI adapted successfully to the changing technologies, Transitron and Hughes did not, with both firms dropping out of the top ten transistor firms by the mid-1960s.

Finally, in the late 1960s the continuing development and refinement of the metal-oxide-semiconductor (MOS) technology led to the adoption of MOS in major commercial data processing applications and provided perhaps the best opportunity ever offered in the industry for potential entrants. (See appendix A for definitions of MOS and bipolar technology.) By 1978 Intel, American Microsystems, Mostek, and other firms entering the industry in the late 1960s and early 1970s, had concentrated on the production of MOS products and risen to positions among the top firms in the industry.

New waves of technology allow new firms to enter and prosper. With the advent of the silicon transistor in 1954 and the planar process in the early 1960s, established firms had no choice but to master the new trends because they obsoleted the older processes. The MOS product cycle was somewhat different in that bipolar technology was, and still is, viable and

The Environment and Technology

was by no means obsoleted by MOS technology. Motorola, TI, and particularly Fairchild, had rapidly expanding bipolar businesses during the MOS take-off.

MOS grew out of the development of field-effect transistors in the early 1960s. General Microelectronics (GMe), a 1963 spinoff from Fairchild, was one of the first firms to attempt to exploit the development of MOS but soon ran into financial difficulties, as the cost and time for this development had been grossly underestimated. Philco-Ford, which bought GMe's facilities, also lost a lot of money in MOS development (Kraus 1973, pp. 51-52).

Not until 1970, spurred by its uses in calculators and computer main memory, did MOS technology really take off. This growth in demand coincides with the growth in total semiconductor shipments during the early 1970s shown in figure 2-3. This suggests that most of the firms entering in the peak before this growth would be firms specializing in MOS technology. Confirming this supposition, in 1973 approximately 85 percent of the sales of firms established in 1966 or after were concentrated in MOS technology while only 35 percent of the sales of firms established before 1966 were in MOS, the remainder being concentrated in bipolar. This distinction became less significant toward 1978 as both new and old firms broadened their product lines to include both bipolar and MOS products. The MOS share of newer firms fell to 70 percent while the MOS share of older firms rose to 40 percent.[1] The importance of a new product cycle for new entrants is emphasized when it is noted that bipolar sales also increased during the early 1970s. Although the future expectations in the late 1960s for both bipolar and MOS sales and applications were clearly optimistic, new entrants in general chose to go with MOS.

Other features of the semiconductor-industry environment have contributed to the pattern of entry and subsequent interaction among entry, growth, and technological change. One important feature has been the supply and mobility of skilled personnel. The labor force has been very mobile and if technical people were dissatisfied, they could easily move to another firm. This labor mobility facilitates new entry by making it easier to recruit workers for a new firm. Much of the new entry, furthermore, has taken the form of groups of skilled personnel spinning off from established firms. Labor mobility and new entry were particularly frenzied during the 1960s and early 1970s in California's "silicon valley," located south of San Francisco. Here is found the headquarters of Fairchild's semiconductor operations, and a large portion of the new entrants in the industry were small firms composed of former Fairchild employees.

The availability of capital is another important factor affecting the conditions of entry. This is especially true of *venture capital*, a highly speculative capital invested in either a new company or a relatively young and untried company. The availability of financing is tied to the general state of the economy. The venture-capital market thus tends to be volatile. During the late 1950s and early 1960s, capital was readily available to finance ventures, and this availability contributed to the pattern of entry shown in figures 2-2 and 2-3. In the past two decades, the high periods for venture capital were 1960-1961 (when capital was not available for all types of ventures), 1968-1969 (when capital went primarily to the semiconductor industry), and 1978-1979 (when capital was more readily available for growing firms than for new ventures) (interview with venture capitalists).[2] The three low periods were 1963-1964, 1971-1972, and 1974-1975. Thus, much of the sharp decline in entry in the 1970s may be due to the difficulty of raising venture captial.

An additional consideration contributing to the decline in entry has been rising capital requirements. As circuit density has increased, semiconductor technology has become even more complex. Reflecting this increased complexity, the cost of an efficient wafer-fabrication facility has tripled in the last five years and is expected to double again by 1985 (*Wall Street Journal* 27 April 1979; *Business Week* 3 December 1979). The cost of the equipment needed to set up a wafer-fabrication facility is expected to increase from $10 million in 1979 to $34 million in the mid-1980s (*Business Week* 3 December 1979, p. 80). All venture capitalists that were interviewed agreed that the start-up costs for a new semiconductor firm are now relatively high. Consequently, they think that recently raised venture captial will more likely finance the expansion of small existing firms rather than the start-up of prospective entrants.

End Uses of Semiconductor Devices and Sensitivity of the Industry to Business Cycles

As the invention of new processes permitted increased development and miniaturization of the integrated circuit, more applications for the product were discovered. Initially, the military was the major end user of semiconductor products, and the government generously supported research by means of R&D and procurement contracts. These research funds and the availability of an initial market with premium prices encouraged entry into the semiconductor field (Finan 1975, p. 11). Other government policies, which resulted in liberal licensing and second sourcing, facilitated the diffusion of technology and helped the industry grow and develop (Utterback and Murray 1977).

An early nongovernment application penetrated by the semiconductor firms was industrial equipment including computers. A more recent and rapidly expanding area of penetration has been consumer applications such as digital calculators, television sets, watches, and children's toys and games (Kraus 1973, p. 151). Table 2-1 shows the proportions of sales to end uses categorized as computer, consumer, military, and industrial. This table clearly shows that consumer and industrial end uses have increased dramatically in importance, while the military's proportion has decreased from 50 to 10 percent and computers have maintained a steady 30 percent of total semiconductor sales.

Although there are a large number of different digital integrated-circuit products, considerable product standardization also exists. At the most aggregate level, digital devices can be classified as either logic circuits, memories, microprocessors, or specialty circuits. Figure 2-4 illustrates the family tree of semiconductor products. Much of the analysis in this study focuses on digital products. Linear integrated circuits and discrete devices are largely excluded. Digital logic circuits perform arithmetic functions, memories store data, and microprocessors contain the central processing unit of a digital computer in a single integrated-circuit device. Specialty circuits combine logic and memory circuits for a specialized application such as in digital watches, calculators, or video games.

Within the logic-circuit category are families of electrically compatible devices. For instance, a logic gate and a flip-flop would be part of a family. Within the memory category are devices serving different purposes such as random-access memories (RAMs), read-only memories (ROMs), programmable read-only memories (PROMs), and erasable PROMs (EPROMs). Microprocessors are generally classified according to the length of "word," such as 4-bit, 8-bit, or 16-bit, that the circuit is designed to use in performing computer operations. (See chapter 4 and appendix A for further discussion of these products.)

**Table 2-1
Distribution of U.S. Semiconductor Sales by End Use**
(*percent*)

End Use	1960	1968	1974	1979
Computer	30.0	35.0	28.6	30.0
Consumer	5.0	10.0	23.8	27.5
Military	50.0	35.0	14.3	10.0
Industrial	15.0	20.0	33.3	37.5
Total value (millions of dollars)	560	1,211	5,400	10,500

Sources: Data for 1960 and 1968 from Texas Instruments estimate as reported in William P. Finan, *The International Transfer of Semiconductor Technology Through U.S.-Based Firms*, Working Paper No. 118 (Washington, D.C.: NBER Inc., December 1975).
Data for 1974 and 1979 from Dataquest.

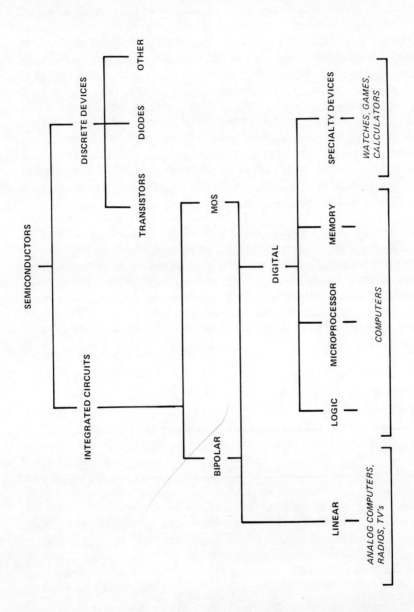

Note: Italic type denotes types of end use.
Figure 2-4. Family Tree of Semiconductor Technology and End Uses

As discussed in chapter 4, when a new product or new generation of a product is introduced, several firms typically vie to get their proprietary design accepted by a large number of customers and thereby become the "industry standard." Other firms will also produce these designs as second sources. In those cases where no clear winner emerges as the industry standard, two or three proprietary designs may remain. However, multiple sources remain for each design. Thus, although a large number of products are in the industry, the typical situation after the initial stage of a new product is that a number of firms will produce one or a few designs for each product. Thus, the large number of products arises more from a large variety of end-use needs than from product differentiation among firms.

Since the demand for semiconductor devices is derived from the demand for final products such as computers, the industry responds very sensitively to aggregate cycles in business activity. If the demand for electronic equipment declines by a certain percentage, for instance, the equipment manufacturer that already has an inventory of semiconductor devices will cut back purchases of semiconductor devices by a much larger percentage. As a result, the swings in output and sales of the semiconductor industry tend to be much larger than for manufacturing as a whole. Declining output and excess capacity occurred during the recessions of 1960-1961, 1969-1971, and 1974-1975, (Webbink 1977, pp. 117-119). Unexpectedly strong demand in 1979, when the start of a recession had been widely forecast, has been blamed for much of the shortage of U.S. firms' semiconductor capacity in that year, making gains by foreign firms easier (*Business Week* 3 December 1979).

Concentration, Foreign Competition, and Vertical Integration

Table 2-2 shows the concentration of U.S. domestic semiconductors shipments in 1957, 1965, and 1972. As seen in the table, concentration has declined slightly over time. With a four-firm concentration ratio of greater than 50 percent, the semiconductor industry is more concentrated than roughly two-thirds of all four-digit manufacturing industries (Webbink 1977, p. 19). However, this moderately high level of concentration masks the substantial turnover in the positions of the leading firms and the diversity of new products introduced in the industry.

A distinction is usually made between merchant semiconductor sales and captive production. *Merchant sales* are those sales made to users outside the firm. *Captive production* refers to the use of the semiconductors produced by a firm in manufacturing its own end-use equipment. Some firms such as TI and Motorola produce semiconductors for both merchant and captive uses. Other firms such as Intel and Fairchild produce almost en-

Table 2-2
Concentration of U.S. Domestic Semiconductor Shipments: 1957, 1965, and 1972

Number of Companies	Percent of Total U.S. Shipments		
	1957	1965	1972
All semiconductors			
4 largest companies	51	50	50
8 largest companies	71	77	66
20 largest companies	97	90	81
50 largest companies	100	96	96
All companies	100	100	100
Integrated circuits			
4 largest companies		69	57
8 largest companies		91	73
20 largest companies		99	91
50 largest companies		100	100
All companies		100	100

Source: U.S. Department of Commerce, *A Report on the U.S. Semiconductor Industry* (Washington, D.C.: U.S. Government Printing Office, September 1979), p. 41.
Note: Data include value of production in captive facilities.

tirely for merchant sales. Two examples of firms that manufacture semiconductors for in-house use only are IBM and Western Electric. Captive suppliers are estimated to account for about 30 percent of total semiconductor production by U.S. firms (*Status* 1979, p. 6-1).

This study primarily concerns those firms that sell a major portion of their semiconductor products to outside end users. Table 2-3 shows the ranking by sales of such firms in the United States for the years 1955, 1960, 1965, and 1975. The latter years reflect the new entry and rapid growth of several firms such as Intel and National Semiconductor (National). By 1965, Sylvania and Westinghouse, both old vacuum-tube companies, have dropped off the list of leading firms. General Electric, also an old vacuum-tube company, does not appear on the list in 1975. The exit of large vacuum-tube companies from the U.S. market is in marked contrast to the survival of these firms in Europe and Japan.

The United States has long been the leader in semiconductor innovation and production. However, U.S. firms are now facing considerable foreign competition in some advanced product lines. In particular, Japanese firms have made inroads into LSI circuits. Table 2-4 shows the world sales of major international semiconductor manufacturers in 1972, 1974, and 1976. As shown in the table roughly half of the largest firms worldwide are U.S. firms.

Table 2-3
Leading U.S. Semiconductor Manufacturers: 1955-1975

1955 Transistors	1960 Semiconductors	1965 Semiconductors	1975 Integrated Circuits
Hughes	Texas Instruments	Texas Instruments	Texas Instruments
Transitron	Transitron	Motorola	Fairchild
Philco	Philco	Fairchild	National Semiconductor
Sylvania	General Electric	General Instrument	Intel
Texas Instruments	RCA	General Electric	Motorola
General Electric	Motorola	RCA	Rockwell
RCA	Clevite	Sprague	General Instrument
Westinghouse	Fairchild	Philco-Ford	RCA
Motorola	Hughes	Transitron	Signetics (Philips)
Clevite	Sylvania	Raytheon	American Microsystems

Source: I.M. Mackintosh, "Large-Scale Integration: Intercontinental Aspects," *IEEE Spectrum*, June 1978, p. 54. Reprinted with permission.

The historical development of European and Japanese firms has been significantly different from that of American firms. (Tilton, 1971, chapters 5 and 6; Webbink, 1977). In the initial stages of the industry, much of the demand for the newest and most advanced products in the United States came from the military and the space program. By contrast, government demand for semiconductor devices was almost nonexistent in Japan and Europe. The semiconductor industries in these countries could serve only demand from industrial and consumer users and consequently lagged behind the U.S. industry in developing advanced semiconductor technology.

However, now that the bulk of world sales consists of both industrial and consumer end uses, foreign firms are no longer at an inherent disadvantage and are rapidly catching up. Foreign firms, sometimes with support from their government, have expanded their integrated-circuit development activities. They also have been acquiring U.S. manufacturers.

The initial semiconductor producers in Europe and Japan in the 1950s were large, diversified manufacturers of vacuum tubes, as was the case in the United States. However, unlike the situation in the United States, these large, vertically integrated and diversified companies have remained the dominant suppliers in their countries. This dominance is due in part to the emphasis in Europe and Japan on development rather than innovation and the lack of new (small) firms entering the industry in these countries.

Semiconductors constitute a major component of electronic products and systems, and vertical integration has become an increasingly important characteristic of the industry. Vertical integration has resulted both from

Table 2-4
Estimated World Sales of Semiconductors by Major International Semiconductor Firms: 1972, 1974, and 1976
(*millions of dollars*)

Company	Country	Estimated World Sales		
		1972[a]	1974	1976
Texas Instruments	United States	405	652	655
Motorola	United States	310	482	462
IBM	United States	275	NA	NA
Western Electric	United States	195	NA	NA
Philips	Netherlands	175	277	260
Fairchild	United States	165	324	307
Toshiba	Japan	145	170	233
Hitachi	Japan	135	195	240
ITT	United States	95	160	144
RCA	United States	80	NA	182
Siemens	Germany	80	142	NA
SGS-ATES	Italy	75	78	NA
National Semiconductor	United States	75	204	263
Mitsubishi	Japan	70	80	94
Matsushita	Japan	70	150	254
General Electric	United States	60	114	113
Secosem	France	50	48	NA
Signetics	United States	50	121	125
Intel	United States	45	115	147
Fujitsu	Japan	40	NA	NA
Nippon	Japan	NA	120	343
Telefunken	Germany	NA	37	NA
General Instrument	United States	NA	63	106
Other Companies		855	1,841	1,834
Total		3,450	5,373	5,762

Source: U.S. Department of Commerce, *A Report on the U.S. Semiconductor Industry* (Washington, D.C.: U.S. Government Printing Office, September 1979), p. 89. Estimates for 1972 based on Commerce Department data and other published sources. Estimates for 1974 and 1976 obtained from Semiconductor Industry Services Volume 2, appendix B, table B1 prepared by Dataquest.

NA = Not available.

[a]Includes internal sales (captive production) for the companies' own use.

semiconductor firms moving forward into end products such as computers and telecommunications systems and from makers of electronic products or systems integrating backward into semiconductor components. In addition, a few semiconductor manufacturers have integrated backward into silicon purification and the growing of silicon crystals.

Incentives for semiconductor firms to integrate forward include the more stable pricing patterns and larger profit margins of the end

The Environment and Technology

product or system (interview with executive of integrated-semiconductor firm). Some semiconductor manufacturers have also allowed themselves to be acquired by an end user to obtain access to additional capital. Yet for many of the semiconductor firms that tried, forward vertical integration was a failure, often because they did not understand the needs of the end users or their customers or because they lacked the necessary distribution network. For example, Intel and Mostek, two of the leading innovators in the semiconductor industry, both failed with their watches and Rockwell failed with its pocket calculator. Only TI and Motorola have succeeded so far in obtaining a major portion of their sales from end products, and the profitability of TI's consumer electronic products is in question (*Fortune*, 3 December 1979).

Incentives for equipment manufacturers to integrate backward into semiconductors include obtaining better control over design specifications and product quality, as well as protecting their end-product markets against forward integration by semiconductor manufacturers (interview with executive of integrated-semiconductor firm). Many manufacturers of electric and electronic equipment have attempted to integrate backward but have often failed, usually because of an inability to keep on the frontier of semiconductor technology (Webbink 1977, p. 66). Other firms such as IBM, Western Electric, and Hewlett-Packard produce semiconductors only for their own needs and not for merchant sales. Currently, captive suppliers account for approximately 30 percent of U.S.-based semiconductor production (*Status* 1979, chapter 6). Motorola and TI are the only equipment manufacturers that have maintained a leading position in merchant semiconductor sales. Other firms like RCA and General Instrument are also present but to a lesser degree.

In the past, the balance of entry conditions in the semiconductor industry has favored highly innovative and dynamic firms. Large, well-established, and well-financed firms did not have any clear advantage. However, with capital costs rising dramatically, venture capital relatively unavailable until recently, the increasing use of semiconductors in all types of products, and rising foreign competition, the climate appears more suited to entry by existing firms integrating vertically by merger than for new firms entering de novo.

Technology

This section identifies the aspects of semiconductor technology and production that have important implications for the innovative behavior and performance of firms in the industry.

The pervasive fact of life facing a potential innovator of semiconductor products is that the price of an innovative product will decline over time.

This fact of life implies that the timing and quality of an innovation determines its profitability. Prices of new products fall rapidly because competitive pressure eventually forces the innovator to price at or near the industry's falling average-cost curve for the new product. Thus, competition acts as a constraint on prices and potential profits for the innovator. Second sources (firms that produce and sell another firm's design with or without a licensing agreement) that can quickly come to market to compete with a recently innovated product can also reap some profit rewards before prices start to rapidly decline. Market competition between the innovative firms and second-source firms forces prices down and those firms that have better yields (percentage of good circuits produced from a production run) and thus enjoy lower average costs (because of moving down the experience curve earlier and faster) than the industry average, can earn above-average profits. The potential to earn these profits inspires innovators and second sources that enter early in the product life cycle.

Prices decline in a competitive environment if costs decline, demand falls, or supply increases. Declining costs have been the primary reason for declining prices in the semiconductor industry. In the last two decades, costs have declined for two principal reasons: (1) the experience curve and (2) increasing circuit complexity. The experience curve for a specific product arises from the phenomenon found in all industries: the more experience an industry has, the more efficient it becomes. Integrated-circuit costs per circuit have declined on average by 28 percent with each doubling of quantity of integrated circuits produced. The uniqueness of the semiconductor industry lies in the rapidity with which firms have moved down the experience curve, and this rapidity in turn has been stimulated by the ever-increasing circuit complexity (Noyce 1977, p. 67).

Figure 2-5 illustrates the effects on costs of both the experience curve and increasing circuit density for semiconductor memories. For a given product (for example, a 4K or more precisely 4,096-bit memory) costs decline over time. However, even more dramatic cost declines result from new products of greater function density per circuit (for example the 16K memory).

This section examines the technological principles of the semiconductor industry that permit rapid price declines and foster industry innovative behavior. We discuss in turn the technological concept of the transistor, the integrated circuit and its manufacturing process, the experience curve versus yield, and increasing circuit complexity.

The Digital Integrated-Circuit-Transistor Roots

The building block and heart of all integrated-circuit technology (and of the semiconductor industry generally) is the transistor. In appearance, the transistor is a tiny (about one-ten-thousandth to one-millionth of a square inch),

The Environment and Technology   27

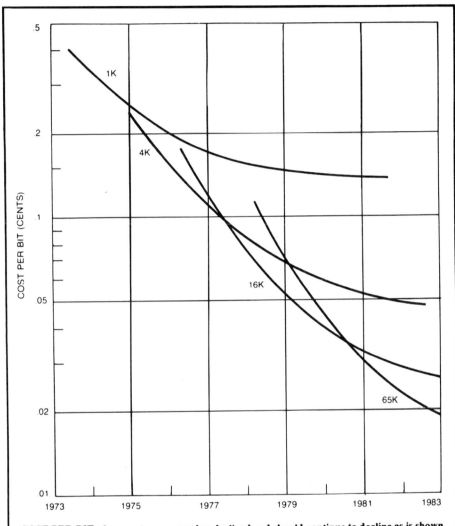

COST PER BIT of computer memory has declined and should continue to decline as is shown here for successive generations of random-access memory circuits capable of handling from 1,024 (1K) to 65,536 (65K) bits of memory. Increasing complexity of successive circuits is primarily responsible for cost reduction, but less complex circuits also continue to decline in cost.

Source: From *Microelectronics*, Robert N. Noyce. Copyright 1977 by Scientific American, Inc. All rights reserved. Reprinted with permission.

Figure 2-5. Estimated Decreases in Computer-Memory Cost: 1973-1983

flat piece of pure silicon crystal to which "impurities" such as antimony and boron are added selectively to different areas of the flat surface. These impurities give different parts of the formerly homogeneous silicon crystal different electrical conducting properties. Antimony impurity atoms imbedded in silicon facilitate the easy conduction of negative charges, and boron impurity atoms facilitate positive-charge conduction. Areas of the silicon inbedded with antimony and boron are called *n* and *p regions*, respectively, because of their respective negative- and positive-conducting properties. After impurities are added to the silicon crystal, the flat surface is covered, again in different regions of the surface, with glass and aluminum. Glass is an insulator and protects the underlying n and p areas and their intersections or junctions. Aluminum is a conductor and the aluminum surface allows n and p regions to make electrical contact with the outside world (that is to other circuit elements outside the transistor).

The transistor has three connections or *leads* to the outside world. When a current is applied to one of the leads (the base lead), a current flows between the two remaining leads. Thus, a transistor acts much like a light switch. When the switch is turned from off to on (hit the base lead with a pulse of current), current flows through the circuit and lights the light (flows between the other two leads).

This switching property of transistors is what makes them attractive as the building blocks of basic computing equipment for computers, calculators, watches, and games. Computers basically manipulate and store numbers and words. In computers all numbers and words are translated into "binary language," where letters, numbers, and words are expressed as groups of ones or zeroes. When the transistor is on, or conducting current, it is usually used as an electrical temporary storage for a binary zero; when off, it stores a binary one.

Since large computers store and manipulate many thousands of ones and zeros simultaneously, many thousands of transistors are needed to make a computer. Furthermore, large computers are expensive to buy or lease and therefore, the faster that the computer can manipulate information, the more cost effective it becomes in government and private offices. Desk calculators, however, make comparatively fewer computations but are driven by batteries, making power drain on the batteries important.

The speed with which two discrete or individual transistors adjacent to each other in a computer can send a signal from one to the other (for example "store this binary one that I am passing over to you") depends on the length of the wires and solder that connect them and the type of transistors involved. Two main types of transistors exist (and thus digital integrated circuits), bipolar and MOS. The important difference between them, when used as a switch, is that the base lead (or light-switch button) is triggered by a current pulse in bipolar and a voltage pulse in MOS.[3] MOS transistors are

The Environment and Technology

smaller in area and use less power (for example, battery drain) than bipolar transistors. Bipolar transistors can turn on and off faster than MOS transistors and are thus attractive for use in the computing and control circuitry of larger computers. Although MOS has been slower than bipolar, the two technologies are now becoming more nearly equal.

The Digital Integrated-Circuit Concept

Manufacturers of silicon transistors in the late 1950s and early 1960s were aware that the lengthy interconnections between transistors slowed down the effective switching speed of even the fastest bipolar transistors. These manufacturers found that if several or several hundred transistors could be built adjacent to each other (that is, *integrated*) in the same piece of silicon crystal rather than connected to each other with solder on a circuit board, the computer transistor-interconnection distances would be a tiny fraction of what they were. The result would be improved speeds. In addition, overall computer reliability could be improved and power drain lowered.

An overriding consideration, however, was cost per function. Would it be cheaper to build a finished computer with the capability of performing a specific array of computing and storing functions at a given speed and power drain, using integrated-circuits or transistors? The semiconductor industry claimed that integrated-circuits would be faster, cheaper per computing function (for example, add two numbers and store the result), more reliable, and less consumptive of power. In short, computer manufacturers in the early to mid-1960s were faced with a very attractive prospect since integrated circuits presented major benefits. Consequently, computer designs relying on discrete transistors were gradually replaced by first-generation integrated-circuit designs, and the digital integrated-circuit products appeared and flourished.[4]

Early digital integrated-circuit designs had fewer than fifty transistors built or grown into the tiny piece of flat silicon. Today's MOS integrated-circuit designs have tens of thousands transistors. Thus today, instead of performing a simple storage or arithmetic function as part of a system of interconnected integrated-circuits making up the computer, one integrated-circuit (the microprocessor) is the entire computer contained in one piece of flat silicon measuring a quarter inch on a side.

Taken individually, early-vintage digital integrated-circuits could not perform as computers; digital integrated-circuits needed to be connected to one another. Therefore, to sell digital intergrated-circuits, semiconductor houses in the 1960s and early 1970s had to provide computer equipment makers with a complete "kit" of electrically compatible parts that could be interconnected into a computer configuration. This resulted in individual digital integrated-circuit products becoming grouped into "families" such as

Resistor-Transistor Logic (RTL), Diode-Transistor Logic (DTL), and the durable Transistor-Transistor Logic (TTL). For example, to be considered a serious competitor by users of TTL, an integrated-circuit manufacturer had to be committed to making most of the parts of a complete kit necessary to build a computer, and not just specific integrated-circuit products that were currently profitable. Consequently, the literature will refer to a TTL family or a DTL family and within these families, individual integrated-circuit parts of different types were bought and sold separately or in package deals.

The Integrated-Circuit-Manufacturing Process

To clearly understand the technological basis for the decline in average cost of production for a specific semiconductor product, it is necessary to examine the integrated-circuit-manufacturing process. Figure 2-6 lists the principal stages in semiconductor manufacturing. Within each stage scores of serial steps must be taken to complete the processing at that stage. Our discussion of the manufacturing process will be an overview limited to an explanation of the major stages and the role these stages play in determining production-cost economics. Since we are focusing on the manufacturing steps rather than product design, we ignore the circuit design and mask preparation stages.

In crystal growing and wafer slicing, a cylindrical ingot of pure silicon crystal is grown in a reactor oven and then sliced into thin silicon wafers. In growing a silicon ingot, a small "seed" crystal is inserted into a container of

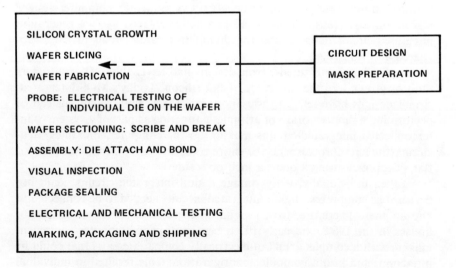

Figure 2-6. Integrated-Circuit-Manufacturing Stages

The Environment and Technology

molten silicon and slowly withdrawn. As the seed is withdrawn from the molten silicon, some of the molten silicon solidifies in crystalline form around the seed crystal forming a cylindrical ingot of adhering silicon crystal. Over the last two decades, ingot and wafer diameters have increased from one to five inches as crystal-growing innovations have permitted larger-diameter wafers to be grown with the necessary crystalline purity and homogeneity.

Wafer fabrication is the stage in which transistors are grown or diffused in the silicon crystal wafers. Each integrated circuit will contain thousands of distinct transistors. The maximum number of integrated circuits produced from each wafer is determined by the size of the wafer and the circuits. For example, on a 3-inch-diameter wafer a maximum of 175 separate integrated circuits, each measuring 1/5 of an inch on the side, can be made.

$$\frac{\text{Area of wafer}}{\text{Area of chip}} = \frac{\pi (d/2)^2}{0.2 \times 0.2} = \frac{\pi (3/2)^2}{0.2 \times 0.2} = 175$$

where

d = wafer diameter in inches,
$\pi (d/2)^2$ = wafer surface area, and
0.2×0.2 = the chip area.

The wafer-fabrication yield or the number of good integrated circuits per wafer will be some percentage of the geometrically determined 175-circuit maximum. Wafer fabrication consists of 5 or 6 major stages in which impurities like antimony and boron (or n- and p-conduction regions) and connectors (aluminum) or protectors (glass) are added to the pure silicon wafer. Growing a p region, for example, is a process of exposing selected silicon-wafer surface areas to an atmosphere of boron atoms and diffusing these atoms throughout the crystal in high-temperature ovens. In each wafer-fabrication stage (say, boron-impurity addition), all 175 circuits on the wafer receive the necessary amount of boron in the necessary areas simultaneously. Furthermore, 20 to 200 wafers may be processed together so that up to $200 \times 175 = 35,000$ integrated circuits will have all their p regions (or positively conducting regions) grown simultaneously. In 6 to 12 sequential stages, involving well over 100 distinct steps, all 35,000 integrated circuits in a production batch or lot are manufactured together.

After wafer fabrication each individual integrated circuit on the wafer is electrically tested. The tester sets a mechanical attachment with tiny needle-like electrical contact points sequentially onto each integrated circuit on the wafer. The probe points send and receive electrical signals to and from the integrated circuits in making the required voltage and current measurements. An ink drop is put on any defective circuits. Then the 3-inch wafer is physically broken into 175 separate rectangular integrated circuits. (The scribe-and-break process is mechanically similar to scratching glass and breaking it along the scratch lines.) Defective or inked units are discarded.

The integrated-circuit now is an isolated, fragile, and thin, flat rectangle measuring one-fifth of an inch on a side. Its chemical and mechanical fragilities are so great that it has to be sealed in a ceramic or plastic case before it can be soldered into a computer-printed circuit board. The case, or *package* as it is called in the industry, must have two properties. The package must be mechanically strong to protect the integrated-circuit (like a turtle shell protects a turtle) and must allow different sections of the integrated circuit to be electrically connected to the outside world (like openings in the turtle shell for legs and nose).

The next stage is die attach, when the bottom of the integrated-circuit is cemented to the package. Then comes the bond stage, when sections of the integrated-circuit are "stitched" with fine wires to the more rugged metal "legs" of the package. Each stitch is made by ultrasonically welding one end of a fine wire to the integrated-circuit and the other end of the fine wire to one of the fourteen or more copper alloy or plated nickel-iron-alloy leads imbedded in the package, which are to be soldered into the computer. Fourteen or more stitches are required for each integrated-circuit. This stitching, or *bonding* as it is called, is a labor-intensive process. Die attach and bond are commonly referred to as the *assembly* section of integrated-circuit manufacturing.

Poor bonds are potential cause of integrated-circuit failures in customer equipment so the bonds and the die themselves are visually inspected on a sample basis before the package is sealed.

After the package is sealed, the finished integrated-circuit is electrically tested and the package is tested for several mechanical integrity properties. Some units fail both tests, but assembly- and test-yield percentages (unlike wafer-fabrication yields) are usually fairly high with 80 percent as a reasonable order of magnitude.

The Experience Curve: The Role of Yields

The experience curve for a given integrated-circuit product stems from yield improvement over the product's life cycle. The greater the number of good circuits at the end of a manufacturing run, the lower the cost per unit will be. A run starting with 3,500 potential circuits will lose many circuits as the run proceeds through successive manufacturing steps. The capital equipment and labor necessary to process a given run are fixed in the short run, so total run costs are relatively constant. With constant run cost, the cost of a good integrated-circuit is simply the run costs divided by the number of good integrated-circuits emerging from the run. Thus, in the short run

The Environment and Technology 33

yields, or the percentage of good units per run, are the fundamental determinants of costs and profits.

At any point in time, most firms have access to current state-of-the-art wafer-fabrication ovens, bonders, and testers. Much of this equipment is supplied by firms that are not vertically integrated with the integrated-circuit manufacturer (Schuyler 1979, p. D4). Therefore, an innovation in process equipment that is available to all manufacturers results in competitive pressures that soon force the adoption of all significant process-equipment innovations.

Assembly and testing areas are not generally a source of major yield losses. While assembly foremen and operators pay close attention to yields, they concentrate particularly on improvement of weaker operators' yield performance. The assembly and testing areas process a variety of products and the assembly and testing yields do not differ much across products. Furthermore, since the same state-of-the-art bonding and testing equipment is available to all integrated-circuit manufacturers, assembly and testing yields will not vary widely among firms.

Therefore, much of the engineering effort in a semiconductor-manufacturing operation focuses on wafer-fabrication yields. Early in the life of a new product, yields may be only one or two good integrated-circuits per wafer. All firms know that if they can get the wafer yields from 1 percent to 10 percent, they cut their wafer-fabrication-area costs per unit by a factor of 10. If second sources are getting 2 percent and the innovator 10 percent yields, the innovator has an opportunity for large profits. Each new integrated-circuit introduced precipitates a race to attain wafer-fabrication yields higher than the current average for the industry. Those firms whose yields are better than average can gain above-average profits or can lower prices and expand market share; those firms whose yields are lower must improve their yields or withdraw.

Improving yields is a matter of applying skilled engineering talent to wafer fabrication and circuit design over a period of time. Circuit designers work with wafer-fabrication engineers to identify product-design modifications that will result in improved yields. The better the engineers and the more time they are allowed to focus on yields, the better the yields and the higher the profits per unit sold. A late-starting second source with superior engineers can catch up to an innovator with a substantial lead time. Similarly, an innovator with superior engineers can maintain its lead.

Therefore, in deciding to make a new product, the integrated-circuit manufacturer looks at its own and its rivals' expected time profile of wafer-fabrication yields. Most firms believe they can achieve the same or better yields than rivals. Firms that expect to achieve industry-average yields may enter anyway for marketing reasons such as to protect market shares or to be a supplier of all the components of a digital integrated-circuit family.

Nonexperience-Curve Yield Improvements

The experience curve describes the time profile of improved yields, declining costs, and declining prices over a specific product life cycle. Other ways to cut manufacturing costs also exist that will simultaneously cut costs over a whole array of integrated-circuit products being manufactured. For instance, yield improvements and cost reductions can result from the purchase of better production equipment such as crystal growers, wafer-fabrication ovens, bonders, and testers. Of course, rivals can also install similar equipment and realize equal improvement in yields. An important innovation of this type has been the drive to push for larger-diameter wafers. Whereas 100 integrated-circuits can be produced on a 2-inch wafer, 400 can be produced on a 4-inch wafer since the area of wafer increases in proportion to the square of its diameter.

Another important cost-cutting strategy, though only indirectly related to yields, has been offshore assembly. Since the assembly-area costs are more labor intensive than wafer fabrication and small, light integrated-circuits can be shipped inexpensively, manual assembly for products produced in large volume is cheaper per unit in lower-wage areas such as Korea and Southeast Asia.

Increasing Function Density: Moore's Law

If rivals are achieving better yields, the integrated-circuit manufacturer can regain profits and market share by inventing a new product. If, hypothetically, wafer-fabrication yields can be improved to the point that later second sources will take two years to catch up, a large profit incentive presents itself to invent a product that is a substantial improvement over existing designs from the customer's vantage point.

In digital integrated-circuits a product improvement usually means packing more computer logic or memory storage into the integrated-circuit. This increase in computer logic or memory storage is accomplished by increasing the function density, or the number of functions that can be performed by the integrated-circuit. Figure 2-5 indicated that successive generations of random-access memories quadrupled the effective memory capacity per integrated-circuit every two years.

The computer designer is concerned with cost per function (assuming adequate speed and acceptable power consumption). If an integrated-circuit manufacturer can sell a new product for less on a cost-per-function basis, the product will "capture" or be designed into emerging generations of computer equipment. In other words, new-product innovation does not occur without regard for the demand for digital integrated-circuits. Many new integrated-circuit designs have been unsuccessful because of lack of

The Environment and Technology   35

customer acceptance: the case of custom LSIs will be examined in chapter 4 on strategy. The expected cost per function of existing designs over the product life of the computer or calculator being designed sets a ceiling-price time path for a new innovation. Computer designers are induced to design with new integrated-circuits if they expect or are led to expect that they will achieve substantially lower cost per function (assuming the same or improved circuit-design features in the new integrated-circuit).

The pace of improvement in function density that has characterized digital integrated-circuits has been stated succinctly as "Moore's Law." In 1965 Gordon Moore of Fairchild predicted that function density would double every year. Moore's Law held true for more than fifteen years. In December 1975 Moore amended his law for the ten years beginning in 1980 by saying that circuit complexity would increase at half the rate (that is, double every two years) of the previous fifteen years (Hogan 1977, p. 32). Circuits presently can contain more than 200,000 elements (Allan 1979, pp. 31-32; Noyce 1977, p. 65).

Conclusion

We have examined the technology of integrated-circuit manufacturing in order to describe the reasons for the rapid pace of product improvement and declines in price that have characterized digital integrated-circuit markets. Two important technologically rooted bases for the cost declines that permitted the observed price declines were described: the experience curve and increasing function density. The experience curve applies to a specific product. Costs per unit drop over time due primarily to improved wafer-fabrication yields. The increasing function density brought about by rapid new-product innovation has been as potent as the experience curve in declining costs per integrated-circuit function. The product-innovation trend is summarized as Moore's Law, which states that the function density of digital integrated-circuits double every year.

Both the drive for improved yields and product innovation derive from the profit incentives available to firms with cost curves lower than the industry average. These profit incentives in turn are available to manufacturers that have the technical knowledge and entrepreneurial ability to exploit them.

Notes

1. The sample of firms used to construct these averages ranged between 33 in 1973 to 36 in 1978, and they were calculated by CRA in November 1978 from firm sales data in *Status*, "A Report on the Integrated Circuit In-

dustry," Integrated Circuit Engineering Corporation, 1973, 1975, 1976, 1978, 1979.

2. According to *Business Week* (3 December 1979, p. 77), $500 million in venture capital has been raised since the capital-gains-tax cut in 1978. Zschau (1978, p. 9), reporting on a major survey of capital availability based on 325 electronic firms, states that "firms founded in 1971-1975 were able to raise on the average *less than 30 percent* as much capital as firms founded during 1966-1970 raised during the 1966-1970 period."

3. Voltage in a wire is analogous to the pressure that forces water to flow through a pipe (for example, in pounds per square inch) while current is analogous to the rate of flow through the pipe (for example, in gallons per hour).

4. As discussed in chapter 1 we focus on digital integrated circuits for much of the analysis of innovative behavior, competition, and performance in subsequent chapters. This focus excludes linear integrated circuits and discrete semiconductor devices. In a linear circuit the output is proportional to the input, in contrast to a digital circuit where the output is either a fixed level, which represents one in binary language, or zero, and therefore nonlinear.

3 Innovative Behavior by Manufacturers of Integrated Circuits

The pace of innovation in integrated circuits has been truly remarkable. An important question is what factors affect innovative behavior in this industry? By answering this question we can then ask how government policy influences the innovative process. This chapter examines the relation between important characteristics of individual firms and their innovative behavior. The first section briefly reviews the framework, which was established in a companion framework volume (CRA, 1980.) The second section is an outline of the innovative record in integrated circuits. The third section summarizes the views of industry executives with respect to the characteristics of innovative firms. The fourth section contains an in-depth discussion of the relative importance of different characteristics based on our case study. A staightforward statistical model based on the material in the second and fourth sections appears in the fifth section.

The conclusions, which are discussed in the final section, are first that certain characteristics of firms are more important than others and that important interrelationships exist among the most important characteristics. These characteristics are the commitment and foresight of top management with respect to innovation, the balance between organizational flexibility and management control, the degree of risk taking, capital availability, and the level of R&D spending.

Second, these characteristics have different effects on different types of innovation. Organizational flexibility and management control are important for both incremental and major innovation. However, the commitment and foresight of top management and a high degree of risk taking are more important for major innovations than for incremental. In contrast, capital availability and R&D spending are more closely related to incremental innovation than major innovation.

Third, some of the characteristics identified in the framework are found to be less crucial than the characteristics mentioned earlier. These less-crucial characteristics include patents, basic research, the technical expertise of marketing staffs, and financial incentives within the firm.

Review of Framework

The general framework of this study hypothesizes that a firm's strategy formulation largely determines its innovative behavior, although uncertainty exists in the relationship between the two due to technical and market risks and the uncertain actions of other firms. Strategy formulation influences and is influenced by the firm's resources and organizational structure. To pursue an innovative product strategy, for instance, a firm must hire skilled personnel and organize their work environment so that they are creative and effective. The presence of these people and the means of firm organization, in turn, affect strategy formulation, which has been described as the process of matching the firm's capabilities with the opportunities it perceives. Observing strategy formulation directly, however, is difficult because much of the process must be confidential to outsiders. Detailed studies of strategy formulation devote considerable resources to analyzing particular firms' actions.

The framework volume identifies a number of characteristics of the firm that are expected to affect strategy formulation and thereby determine innovative behavior. These characteristics relate to the skills and orientation of the firm's personnel, the financial resources the firm has available for innovation, and the firm's organizational structure. Personnel skills and orientation include top management's background and interest in innovation and the technical expertise of R&D, production, and marketing staffs. Financial resources include the level of R&D spending, the level of basic-research effort, capital availability, and the strength of the firm's patent position. Organizational characteristics include the techniques for integrating R&D with production and marketing, organizational flexibility, and the incentive structure within the firm. Finally, the degree of risk taking is an indicator of both financial resources and personnel skills and orientation.

The Innovative Record

Our analysis of innovative behavior uses two alternative measures of innovative output. The first measure is to identify a reasonably complete list of major or key innovations or innovations that represent a technological breakthrough compared with previous practice. The key innovations listed in this section have been identified through consultation with industry experts and through interviews with industry executives.

The second measure of innovative output is the number of semiconductor patents a firm receives. Although some of these patents will apply to major innovations, many more will apply to less-dramatic advances. Therefore patent totals are more a measure of incremental innovation.

Although the distinction between incremental and major innovations refers to the relative importance of particular technological advances, the cumulative effect of incremental innovations is very important and may well outweigh the cumulative effect of major innovations. In fact, many major innovations would likely be viewed as relatively incremental if they simply occurred at a later date. A similar point can be made about the identity of a major innovator. If the particular firm that made the innovation did not do it at that particular time, another firm very likely would have made the advance, either as a major or an incremental innovation, at a later date.

In discussing integrated-circuit innovations it is useful to distinguish among device-structure, process, and circuit-design innovations. Processes are those techniques used in the production of semiconductor devices. Structures are the semiconductor components that are constructed with the processes. Circuit designs interconnect the basic semiconductor components to perform electronic functions such as amplification, switching, and storage of information. Individual circuit designs are usually classified into families based on common design features, processes, and structures.

Tables 3-1 through 3-3 show key device-structure, process, and product-family innovations in digital integrated circuits. (Some device-structure and process innovations shown apply to linear as well as digital circuits.) Brief descriptions of these innovations are contained in appendix A. Several points about these tables are important for clarification. First, in many cases a new device structure defines a new process and vice versa. In tables 3-1 and 3-2 the innovation is categorized as to where the essential breakthrough was made. For instance, the MOS innovations are generally classified as device-structure advances while the Schottky junction is classified as a process advance, although in these cases the structure and process are closely linked. Second, in many cases a device-structure/process innovation makes a product family feasible. For instance, the CMOS transistor made CMOS logic possible, and polysilicon deposition made MOS dynamic random-access memories feasible.

We can also obtain data on semiconductor patents. Tables 3-4 through 3-6 show the timing of semiconductor-device, process and packaging, and digital systems and applications patent applications from 1969 through 1976 for the ten largest merchant integrated-circuit firms and for the ten largest patent holders excluding the largest merchant firms. (As used in this study a "merchant" firm is one that is active in sales of integrated circuits to equipment makers and other customers. In contrast some firms produce only for their own "captive" use.) For clarification, please note that some patents will apply to more than one category and be counted in more than one table. Thus, it is not possible to sum across the tables to obtain a firm's total number of semiconductor patents.

Several features of the patent data are important to note. One is the

Table 3-1
Semiconductor Device Structures

Device	Approximate Introduction	Originator/ Early Advocate
Integrated circuit	1960	Texas Instruments/Fairchild
Diffused resistor	1960	Fairchild
Metal-oxide-semiconductor (MOS) capacitor	1960	RCA
Metal-oxide-semiconductor (MOS) transistor	1962	RCA
Substrate-diffused-collector transistor	1962	Sylvania/Fairchild
Complementary metal-oxide-semiconductor (CMOS) transistor	1963	RCA
MOS resistor	1965	Fairchild/GMe
Deposited-metal resistor	1965	Source varies
Depletion-mode MOS resistor	1969	Mostek
Bipolar-junction field-effect combination	1970	National
Trapped-charge storage	1972	Intel
Bipolar-MOS combination	1972	RCA
Isoplanar, Coplamos et al.	1974	Fairchild/Standard Microsystems
Merged transistor I^2L	1975	IBM/Philips
Vertical MOS (VMOS)	1977	Siliconix/American Microsystems

Source: CRA, *International Technological Competitiveness: Television Receivers and Semiconductors*, prepared for National Science Foundation (Boston, Mass., June 1979). List of innovations developed in conjunction with Ferguson Associates, Cupertino, California.

very large number of patents compared to key innovations identified in tables 3-1 through 3-3. Second, a large amount of patent activity takes place by firms that are not active in merchant integrated-circuit sales. As discussed in chapter 2, this study focuses primarily on the merchant semiconductor firms. However, the greater representation of nonmerchant firms in the patent statistics rather than in the key innovation lists is worth discussing briefly. For those firms that are captive semiconductor manufacturers, the timing of a technological advance is largely determined by their end-use needs rather than by any need to gain semiconductor sales. For instance, IBM used hybrid integrated circuits in its third generation of computers because the superior monolithic integrated-circuit technology could not be developed in time to meet its targets for equipment development (Bloch 1979, p. 91493).[1] Decisions such as this will result in captive firms achieving fewer key innovations per dollar of resources spent. Another factor that may contribute to patent totals of the large nonmerchant firms is that some of these firms such as IBM, Western Electric, RCA, and General Electric conduct basic research on semiconductors (see also pp. 64-65). Many patents

Innovative Behavior by Manufacturers

Table 3-2
Semiconductor Processes

Process	Approximate Introduction	Originator/ Early Advocate
Resistive metal deposition	1960	Source varies
Ultrasonic bonding	1961	Fairchild
Annealing for stress relief	1965	Bell Laboratories
Gettering processes	1965	Bell Laboratories
Plastic encapsulation	1965	Fairchild
Direct bonding (bump)	1966	IBM/Western Electric
Nitride-chemical vapor deposition (evp)	1968	General Instrument
Glass-chemical vapor deposition (CVD)	1969	Bell Laboratories
Ion implantation	1970	Mostek
Schottky junction	1970	Texas Instruments
Dielectric isolation	1970	Harris
Sputtering	1973	Veeco
Polysilicon deposition	1973	Intel
Silicon-on-sapphire	1973	RCA
Plasma etching	1974	Standard Telecommunication Labs, England
Ion milling	1975	Veeco
Electron-beam processing	1975	Texas Instruments/IBM
Photolithography projection alignment	1976	Perkin-Elmer
Deep ultraviolet photolithography	1977	Perkin-Elmer

Source: CRA, *International Technological Competitiveness: Television Receivers and Semiconductors*, prepared for National Science Foundation (Boston, Mass., June 1979). List of innovations developed in conjunction with Ferguson Associates, Cupertino, California.

Table 3-3
Key Digital Integrated-Circuit Product Families: 1961-1975

Product Family	Approximate Introduction Date	Originating Firm(s)
RTL (resistor-transistor logic)	1961	Fairchild
DTL (diode-transistor logic)	1962	Signetics
ECL (emittor-coupled logic)	1963	Motorola
TTL (transitor-transistor logic)	1964	Sylvania/Texas Instruments
CMOS	1968	RCA
Schottky TTL	1970	Texas Instruments
1K MOS RAMs	1970	Advanced Memory Systems/Intel
Microprocessor	1971	Intel
Bipolar RAMs	1972	Fairchild
4K MOS RAMs	1973	Intel/Mostek/Texas Instruments
16K MOS RAMs	1976	Intel/Mostek

Source: CRA interviews with industry executives and industry trade literature.

Table 3-4
Semiconductor Device Patent Applications: 1969-1976

Firms	1969	1970	1971	1972	1973	1974	1975	1976	Total
Merchant Integrated-Circuit Firms									
Texas Instruments	15	9	27	9	21	37	17	24	159
Fairchild	5	5	2	1	7	10	10	5	45
Motorola	18	31	32	30	25	39	26	27	228
National Semiconductor	4	2	2	4	2	5	6	16	41
Intel	—	2	3	3	5	4	11	9	37
Signetics	2	2	8	3	3	9	5	2	34
RCA	32	52	42	36	50	41	51	61	365
General Instrument	5	7	3	5	1	6	—	—	27
Mostek	—	—	5	—	2	4	4	3	18
AMD	—	—	—	—	—	—	1	—	1
Other Large Patent Holders									
IBM	64	48	52	69	51	53	68	37	442
Bell Laboratories	44	52	43	39	32	30	23	19	282
General Electric	27	32	35	34	19	20	24	22	213
U.S. Navy	38	26	33	21	19	21	21	11	190
Westinghouse	31	19	21	21	24	12	19	18	165
U.S. Philips	23	20	26	25	24	22	26	18	184
Hitachi	8	7	13	16	29	24	28	47	172
Siemens Aktiengesellschaft	8	13	18	19	9	30	25	32	154
Honeywell	12	12	13	7	8	8	11	7	78
Rockwell International	12	17	13	4	6	16	9	18	95

Source: Office of Technology Assessment and Forecast, U.S. Patent and Trademark Office. This table represents device-related semiconductor patents identified by the following classes and subclasses:

Class	Subclass
357	1-92

Note: Data for patents granted through 1978. The application data for 1975 and 1976 will be slightly understated. Patent applications that were filed in those years, yet which have not been awarded as of 1978, are not recorded in the OTAF data.

for these firms could represent inventions that flow from their basic research. These inventions need not necessarily be put into practice as innovations.

A third feature of the patent data is that the U.S. government is among the largest semiconductor patent holders; if the list were extended to the top twenty patent holders, the Army and NASA would join the Navy on the list. Fourth, foreign companies such as Philips, Hitachi, and Siemens are among the top U.S. patent holders; again, if the list were extended, other foreign companies would rank highly. Finally, some firms such as Motorola, National, and Signetics rank higher compared to other merchant integrated-circuit firms in terms of patents than in terms of key innovations. This last contrast between the two measures is explored further in later sections.

Table 3-5
Semiconductor Process and Packaging Patent Applications: 1969-1976

Firms	1969	1970	1971	1972	1973	1974	1975	1976	Total
Merchant Integrated-Circuit Firms									
Texas Instruments	37	31	21	18	30	20	15	15	187
Fairchild	7	13	5	4	7	9	9	6	60
Motorola	15	32	32	28	40	17	19	14	197
National Semiconductor	2	—	—	—	2	3	7	3	17
Intel	—	1	—	1	—	2	4	5	13
Signetics	7	1	5	7	12	12	8	1	53
RCA	26	34	34	28	29	35	41	30	257
General Instrument	2	3	1	1	2	3	—	—	12
Mostek	—	—	—	—	1	—	—	—	1
AMD	—	—	—	—	—	—	—	—	0
Other Large Patent Holders									
IBM	56	53	54	57	60	50	71	64	465
Bell Laboratories	22	44	38	40	23	34	25	27	253
General Electric	31	37	21	20	32	23	38	29	231
Siemens Aktiengesellschaft	29	25	29	12	33	31	25	21	205
U.S. Philips	17	20	27	24	28	25	22	24	187
Hitachi	21	17	30	14	22	29	26	24	183
Westinghouse	19	15	4	11	19	9	5	15	97
Western Electric	5	19	20	17	10	12	13	8	104
Hughes	8	3	7	3	7	10	9	11	58
Sprague	13	2	7	3	7	—	6	6	44

Source: Office of Technology Assessment and Forecast, U.S. Patent and Trademark Office. This table represents process-related semiconductor patents identified by the following classes and subclasses:

Class	Subclass
29	569-591
96	36.2
148	1.5-191, 33-33.b
156	625-662
174	52R-52P
204	192R-192N
250	492A
427	82-99

Note: Data for patents granted through 1978. The application data for 1975 and 1976 will be slightly understated. Patent applications that were filed in those years, yet which have not been awarded as of 1978, are not recorded in the OTAF data.

Characteristics of Innovative Firms: The Views of Industry Executives

In conducting the case study we adopted two approaches to assessing the relative importance for innovation of the characteristics outlined in the first section. In the first approach we collected information on each characteristic and then analyzed the association between particular characteristics and the available data on innovation. The fourth and fifth sections present these results. The second approach was to ask industry executives what characteristics distinguish the most innovative semiconductor firms from other firms in the industry.

In most cases the two approaches agree. The characteristics that appeared to be important from examining the data were also cited by industry executives. The characteristics that the data suggested were less critical for innovation were never cited by the executives. However, in two instances the executives' opinions add important insights. One is the importance of a

Table 3-6
Semiconductor Systems and Applications Patent Applications: 1969-1976

Firms	1969	1970	1971	1972	1973	1974	1975	1976	Total
Merchant Integrated-Circuit Firms									
Texas Instruments	53	35	36	23	34	42	25	24	272
Fairchild	13	15	8	8	8	13	11	10	86
Motorola	15	35	38	29	32	33	23	23	228
National Semiconductor	6	2	1	1	3	5	13	8	39
Intel	1	2	3	·4	2	4	7	6	29
Signetics	9	4	8	8	14	18	12	2	75
RCA	41	55	49	33	42	45	66	40	371
General Instrument	2	6	1	2	1	5	—	—	18
Mostek	—	—	1	—	2	—	—	1	4
AMD	—	—	—	—	—	—	—	1	1
Other Large Patent Holders									
IBM	67	57	58	62	57	49	80	51	481
Bell Laboratories	51	54	49	41	28	33	33	23	312
General Electric	48	43	28	32	43	30	50	32	306
U.S. Philips	36	32	39	23	38	44	29	21	262
Hitachi	32	23	39	20	34	37	42	37	264
Siemens Aktiengesellschaft	27	19	25	20	27	24	34	37	213
Westinghouse	34	18	10	19	25	18	17	14	155
Nippon	20	16	9	6	5	12	8	8	84
Sony	9	6	2	7	12	22	37	14	109
Matsushita	13	12	14	10	12	7	3	7	78

Source: Office of Technology Assessment and Forecast, U.S. Patent and Trademark Office. This table represents process-related semiconductor patents identified by the following classes and subclasses:

Class	Subclass
365	103-105, 174-188, 208, 212, 221, 226-229,
307	200-325

Note: Data for patents granted through 1978. The application data for 1975 and 1976 will be slightly understated. Patent applications that were filed in those years, yet which have not been awarded as of 1978, are not recorded in the OTAF data.

balance between organizational flexibility and management control, which includes the techniques for integrating R&D with production and marketing. Although organizational flexibility appeared to be important based on the record of several large firms that had difficulty keeping pace with smaller, more flexible firms in integrated circuits, we found that neither the data nor the interviews allowed us to distinguish among firms with respect to management techniques. A second insight gained from the interviews was that nonmonetary considerations were important in motivating individuals within the firm.

During our interviews with semiconductor-industry executives we asked the following question: What characteristics (for example, organization, managerial philosophy, resources) distinguish the most innovative semiconductor firms from other firms in the industry? The number of characteristics cited ranged from one to four, with most executives citing two or three characteristics.

Table 3-7 summarizes the distinguishing characteristics cited by executives. The responses must be interpreted carefully because the interviewees did not necessarily refer to the same firms. Nevertheless, the results are interesting. The item mentioned most frequently was top-management characteristics as related to technology. This item was described variously as management commitment to technology, management emphasis on technology, or technical foresight of management. Organizational flexibility was the next most frequently cited characteristic. An important aspect of this flexibility is small firm size, which gives firms the ability to react quickly

**Table 3-7
Distinguishing Characteristic of the Most Innovative Semiconductor Firms**

Summary of Responses by Executives of Nine Semiconductor Firms	
Characteristics	Number of Times Cited
Top-management commitment to technology, technical foresight	5
Organizational flexibility, ability to make timely decisions	4
Management techniques and control	4
Ability to attract creative people, background and experience of key persons	3
R&D spending	2
Access to capital	2
Other: perseverance, strategies, aggressiveness, close links with universities (one each)	4

Source: CRA interviews with industry executives.

to new developments and make timely decisions. Four executives emphasized the importance of management control. One executive described this characteristic as the ability to conduct well-planned, rational R&D programs. Another emphasized the ability of successful firms to push profit-and-loss criteria deep into the organization.

Three interviewees cited the importance of talented individuals. Two described this as the ability of the firm to attract creative people, while one executive emphasized the background and experience of key personnel. Two executives mentioned a high level of R&D spending as an important characteristic. Two other executives cited the importance of access to capital. Finally, several characteristics such as perseverance despite occasional setbacks and close links with universities were mentioned once by an interviewee.

The Relative Importance of Semiconductor-Firm Characteristics for Innovative Behavior

This section discusses the relative importance of characteristics of semiconductor firms in promoting innovation. The information is drawn heavily from the industry trade literature and from responses to interview questions directed at the specific items. Our interviews included questions, for example, about patents, R&D activity, top management's background and allocation of time to innovation, risk taking, management techniques, and incentive systems.

The next three subsections discuss the characteristics that our analysis has shown to have the strongest effect on interfirm differences in innovative behavior. The first subsection discusses top management's role. Next is the balance between organizational flexibility and management control. The third subsection discusses R&D spending and capital availability and risk taking together, since they are closely related. The final subsection discusses characteristics that do not explain a great deal of the variation in innovative behavior among semiconductor firms, including basic research, patents, technical expertise of marketing staff, and financial incentives within the firm.

Top-Management Background and Commitment

One of the most important factors affecting innovative behavior is the experience, insight, and commitment of top management with respect to technology. In many instances in the semiconductor industry, technological entrepreneurs have left existing firms to start their own new firm based on an innovation. In other cases, technological entrepreneurs have risen through the ranks of existing firms to top positions.

Profiles of some of the key individuals heading integrated-circuit firms illustrate the relation of management background and commitment to innovation. In 1957 eight scientists and engineers left Shockley Transistor to form the semiconductor division of Fairchild Camera and Instrument Corporation (Kraus 1973, p. 32). In the early 1960s Fairchild was probably the most innovative semiconductor firm by far, although internal and marketing problems prevented it from capitalizing on many of its innovations. These problems are discussed further in the following section.

Four of the original eight Fairchild founders left in 1961 to start Amelco, a Teledyne subsidiary. Two of the remaining men, Robert Noyce and Gordon Moore, formed the highly successful Intel Corporation in 1968. Jean Hoerni, part of the Amelco venture, eventually founded Intersil in 1967, which has been reasonably successful. At Fairchild, Hoerni invented the planar process, which has been the basic process used in semiconductor production since that time.

The Intel success story is widely known and provides a good illustration of the relation of management background to innovative behavior. As indicated in tables 3-1 through 3-3, Intel has introduced a number of major innovations since its founding. Robert Noyce, who has a Ph.D. in physics from MIT, worked at Philco before joining the Shockley group. At Philco he invented the surface-barrier transistor (Kraus 1973, pp. 33-34). At Fairchild he invented the planar integrated circuit, widely regarded as the origin of today's integrated-circuit industry.[2] Gordon Moore received a Ph.D. in chemistry from Cal Tech. In addition to his commercial and research successes, he is well known for Moore's Law, in which he accurately predicted in the early 1960s that the density of integrated circuits would double each year. In forming Intel, Noyce and Moore recruited Andrew Grove as the head of operations. Grove, who had been at Fairchild and teaching part-time at Berkeley, is now the president of Intel.

Other top executives have also played important roles in innovation through their insights into semiconductor technology. At Motorola, C. Lester Hogan headed the semiconductor division until he was hired as chairman of Fairchild in 1968. Hogan has a Ph.D. in physics from Lehigh University and was a professor at Harvard for several years before joining Motorola in 1958. Under his leadership, Motorola developed its only major innovation, emitter-coupled logic (ECL), and achieved a reputation for excellence in production engineering. Very early in the development of integrated circuits, Patrick Haggerty, now retired from the chairmanship of TI, predicted a trend whereby electronic controls would replace other modes of control (Noyce 1977). As is true of most other top semiconductor executives, TI's current top executives have technical backgrounds. Mark

Shepherd, the chairman, and Fred Bucy, the president, have master's degrees in electrical engineering and physics, respectively. Both men rose to their current positions through the operations and manufacturing portions of the company.

Although most top semiconductor executives have technical backgrounds, the nature and orientation of the background does vary and in many cases is indicative of the firm's strategies. At the research extreme are the top people at Intel. At the production-oriented extreme is Charles Sporck, chief executive of National. Under his close attention National strives relentlessly to cut production costs. At the same time however, the firm has been at the forefront of semiconductor technology in several areas. In table 3-1, for instance, National is credited with the bipolar-junction field-effect combination. It has also been one of the leaders in linear integrated-circuit technology, an area that is excluded from much of our analysis. In addition, National was one of the early firms in the microprocessor race discussed in chapter 4.

At the marketing extreme is Jerry Sanders, chief executive of Advanced Micro Devices (AMD) and an engineer who formerly ran Fairchild's marketing efforts. AMD has recently made a strong move into proprietary designs, which result from a firm's own R&D effort in contrast to second-source production, which involves supplying another firm's design. AMD has also been one of the fastest-growing semiconductor firms. In its promotions and facilities AMD projects a first-class image, in contrast to National's "bare pipe-racks" image. Sanders' office, for instance, is very plush and large enough to hold several of National's top officers, who share a large room with their desks separated by partitions.

The amount of time that the top executive devotes to innovation-related matters also appears to be a good indicator of technological commitment and the firm's relative degree of innovation. Management allocation of time and attention among various functions provides a clear view of the most pressing and important issues to the company. For companies that are at the cutting edge of technology and new markets, strategic planning and innovation require a major portion of time at the senior-executive level. For companies that place less emphasis on innovation and more emphasis on other strategies, senior executives will spend relatively less time on innovation and more on other activities.

Table 3-8 summarizes information on the allocation of top executives' time to innovation and compares this with the record of innovation for the companies surveyed. For the two most highly diversified companies in the sample, the data collected applied to the division head in charge of semiconductors. Otherwise the data are for the chief executive officer of the company. The percentage of time indicated is for innovation and strategic planning closely related to innovation. Surprisingly, the responses fall neatly in-

Table 3-8
Proportion of Top Executives' Time Spent on Matters Related to Innovation and Number of Key Innovations

Percent of Executives' Time	Number of Companies	Key Innovations 1968-1977
≥30	3	10
15-25	3	6
10	3	1

Source: CRA interviews with industry executives. Innovation data from tables 3-1 through 3-3.

to three groups. At the low end, three companies indicated that approximately 10 percent of the top executive's time was spent on matters related to innovation. At the high end, two companies indicated figures of 30 percent and a third company indicated a figure in that range but probably somewhat higher. In between, three companies were at approximately 24 percent, 22.5 percent, and 15-20 percent. The number of major innovations attributed to these companies in the second section is also shown. A clear, positive relationship between top-management time and major innovations is evident.

We would be incorrect to conclude, however, that a company can increase its innovative performance merely by having its top executives devote more time to innovation. Rather, the time devoted to innovation reflects a balance between the abilities (and interests) of top management and the needs of the organization. For some executives and firms more time spent on innovation would be less productive than other activities. In other words, the time devoted to innovation can be viewed as an equilibrium arrived at by maximizing the benefits from that activity given the individual's talents and interests and the firm's strategies and opportunities.

The interest and commitment of top management to innovation is also closely related to the firm's risk taking, which is discussed in more detail later. This close relationship can also be viewed as following from the balance between the talents of top management and the needs of the organization. Some individuals will be better suited to undertaking higher levels of risk due to their experience and insights into the technology and market. The relationship between top-management time devoted to innovation and risk taking is illustrated in table 3-9, which summarizes interview responses. Two of the three executives who indicated their firm undertook risks greater than the industry average also devoted 30 percent or more of their time to innovation-related matters. The third greater-than-average-risk firm fell in the middle category for top-management time. The remaining five firms indicated an average level of risk.

Table 3-9
Proportion of Top Executives' Time Spent on Matters Related to Innovation and Degree of Risk Taking

Percent of Executives' Time	Number of Companies	Number of Companies with Greater-than-Average Degree of Risk Taking[a]
≥30	3	2
15-25	2	1
10	3	0

Source: CRA interviews with industry executives.
[a]Usable responses to the risk-taking questions were obtained from only eight companies.

To summarize, then, the background and talents of top management give particularly useful insight into innovative behavior and competitive strategies in general. For example, at the research extreme are Intel's top executives. Reflecting their interests, Intel's corporate objective is to distinguish itself through technical excellence.

The amount of time that top executives devote to innovation-related matters also appears to be a good indicator of technological commitment and relative degree of innovation. In addition, the interest and commitment of top management to innovation is closely related to the firm's relative level of risk taking. Both the time allocation and degree of risk taking reflect a balance between the abilities and interests of the top individuals and the firm's strategies and opportunities.

The Balance between Organizational Flexibility and Management Control

This section explores the relationship between innovative behavior and two important aspects of organizational structure: the flexibility with which the organization can respond to external events, particularly changes in technology, and the control systems that enable the organization to assess and direct its activities. Rapid response to external developments is critical for a firm's success in an environment such as integrated circuits where a number of aggressive firms are in business and technology changes rapidly, because otherwise a firm can be left with little or no share of new product's sales. However, in order to make speedy investment decisions and mobilize resources, management needs an adequate picture of the firm's operations.

Innovative Behavior by Manufacturers 51

The importance of suitable management control is compounded by the multitude of opportunities facing integrated-circuit manufacturers. Companies must choose some opportunities to pursue rapidly, some more slowly, and some perhaps not at all.

Three general types of firm organizations exist in the semiconductor industry. The first type of firm is the small, autonomously managed firm. Some of those firms have remained small, others have grown large, and still others have left the industry. Intel is an example of a small firm growing large in this class of firms. The second type of firm is the autonomously managed division of a larger firm. Until its recent acquisition by Schlumberger, a French company, Fairchild Semiconductor was an example of this second type of firm, being the largest division of Fairchild Camera and Instrument Corporation. TI's semiconductor division also fits into this class of firms. A third class of firms is composed of divisions of large conglomerates that do not enjoy complete autonomy with respect to management decisions. Examples are such firms as RCA, General Electric, and Sylvania.

We now summarize some highlights from the histories of the nonautonomous divisions of conglomerates such as Philco-Ford, Sylvania, RCA, Westinghouse, and General Electric. The highlights suggest that lack of organizational flexibility played an important role in the lack of success of the firms discussed. However, there is little evidence on the reasons for the lack of success of some electronic giants such as General Electric and Westinghouse in the integrated-circuit industry. The fact that they were participants at one time and later withdrew points indirectly to the hypothesis that top corporate management in these firms may have been too inflexible in the management of their semiconductor division. A competing hypothesis, discussed in chapter 5, is that these firms withdrew because the integrated-circuit industry was highly competitive. The large firms could count on buying integrated circuits from the more specialized firms at competitive prices. The large firms might also be able to earn greater or more stable profits by employing their assets in more sheltered or less-risky industries than integrated circuits.

Philco-Ford entered the integrated-circuit business in February 1966 with the purchase of GMe. Philco's semiconductor division was researching semiconductors as early as 1951 and had a large, highly automated transistor operation in place in the 1950s (Golding 1971, pp. 167-168). After Philco's 1962 merger with the Ford Motor Company, Philco-Ford withdrew from transistor production in 1963. Philco-Ford rejected GMe proposals to organize the semiconductor operations as a separate subsidiary. Problems also arose from Philco-Ford's refusal to treat the semiconductor division's salary and incentive structures differently from other Ford divisions and from the added layers of investment decision making that being a division of a large company required (interview with industry consultant). Within months Howard Bobb and Warren Weaver, key GMe men, left to form American Microsystems Inc. (AMI) (Kraus 1973, p. 49). Electronic Arrays was also formed later by former GMe employees. Philco-Ford suffered losses through the late 1960s and withdrew from the industry in 1971 (Kraus 1973, p. 62).

Sylvania, a subsidiary of General Telephone and Electronics, is another large firm for which there is some evidence of organizational problems. Sylvania, a vacuum-tube manufacturer, began production of transistors in 1953 (Tilton 1971, p. 52). In the early 1960s Sylvania led the development of TTL. (This story is related in more detail in chapter 4.) According to *Business Week* (10 October 1970), the company had manufacturing problems with the introduction of TTL and fell behind in deliveries. TI entered with circuits encased in plastic rather than in the metal or ceramic demanded by the military and became the TTL leader. Sylvania eventually dropped from being the TTL leader in the early 1960s to near last place (*Electronics* 12 October 1970).

Although the evidence is indirect, lack of organizational flexibility may have been partly responsible for Sylvania's failure in TTL. An article in *Electronics* (12 October 1970) attributes the Sylvania failure to top management's refusal to fund advanced development programs. The article also claimed that a series of repressive administrative decisions may have hurt the firm in other areas.

RCA, a large tube manufacturer, has survived in integrated circuits, but it has had many problems implementing advances in technology in the semiconductor industry. One account (*Electronics* 2 March 1970) suggests that RCA has suffered from poor integration of R&D and marketing. William Hittinger, formerly the general manager of RCA's Solid State Division, also states that RCA has had problems in translating technology into marketable products (*Electronics* 28 September 1970). Because RCA's R&D labs have a large degree of independence, they have often developed ideas the company could not easily produce and also produced many marketable ideas that management simply could not appreciate. In the early 1960s RCA was one of the first firms to develop MOS but did not put the technology into production. Then in the late 1960s and early 1970s, after MOS became popular, it introduced CMOS, but again the market potential was small compared to other new-product lines.

The more-general question that arises from the histories of the firms owned by conglomerates and the large but autonomously managed firms is how have some firms succeeded at both innovation and day-to-day production operations in the rapidly changing environment of the integrated-circuit marketplace? The process appears to have been one of staying on top of the technology while simultaneously evolving the management techniques to cope with rapid change. The necessary flexibility seems to be a matter of the right management setting up the right level of control in the firm's policies, systems, and procedures. The histories of firms like RCA, Philco-Ford, General Electric, Westinghouse, and Sylvania point to the possibility that these firms were too tightly controlled and thus too inflexible. Firms like Fairchild, Signetics, Motorola, Intel, and National demonstrate by their success that they have an acceptable balance between organizational flexibility and management control. Observers in the late 1960s, for example, attributed the success of Motorola and Signetics in semiconductors to two factors: (1) organization as an autonomous division within the corporation and (2) top semiconductor executives who were effective in "handling the erratic geniuses who make the business run" as well as in dealing with corporate management with respect to the needs of the semiconductor business (*Business Week* 26 July 1969).

The successful firms have not been without problems, however. Fairchild, for instance, is said to have had at one time the best R&D in the industry but never got it into production (interview with industry consultant). Industry observers attribute much of Fairchild's problem at this time to poor integration of R&D with production. In particular, since the production units at Fairchild had profit-and-loss (P&L) responsibility, they were reluctant to introduce new products which, initially at least, were always less profitable than older products (interviews with industry executives). In late 1971 when Fairchild made a number of changes in organizational structure, the central R&D lab, which had been in a separate group, was moved into the semiconductor group (*Electronics* 6 December 1971). Fairchild's technological orientation and incentive structure also contributed to poor integration of R&D with production. Fairchild's founders were very committed to technology *per se*. In contrast, the orientation of the top semiconductor executive at Motorola was toward "product engineering and getting the product to the customer" (*Electronics* 19 August 1968, pp. 45-47). Incentives for engineers at Motorola were oriented toward working in production rather than in R&D. However, at Fairchild engineers were reluctant to work in production.

Besides the lack of adequate systems, procedures, and controls with respect to the integration of R&D and production, in the late 1960s Fairchild lacked the procedures and controls necessary to handle its interface with its customers in an adequately responsive manner. Fairchild's internal paperwork and production-scheduling systems made it very difficult for customers to get good responses from Fairchild both before and after their order was placed. Customer requests for quotes were mishandled and requests for delivery information either produced no information or unreliable information (interview with former industry employee).

Many of the newer semiconductor firms have learned from the Fairchild experience with respect to management controls. Most firms decentralize their R&D into product groups. Only the most fundamental research is done centrally. In addition to decentralizing R&D, most of the newer firms use a matrix organization to integrate marketing and production. A matrix organization requires employees to report in two distinct directions or lines of authority. For instance, the employee will report to a marketing supervisor and also to a production supervisor. Among other advantages, this type of organization helps make the firm more responsive to customers. For example, National organizes its operations around a product-group matrix structure. Each of its product groups is run as an independent P&L center. Examples of P&L centers for logic products are CMOS, Interface, TTL, and Low-Power Schottky TTL. Each P&L center usually has its own design engineering, wafer fabrication, production control, and product marketing (which covers pricing and product specification). Therefore, each P&L group has a direct incentive to integrate and harmonize the activities of all stages and functions (interview with industry executives).

TI has long advocated the importance of management skills in implementing new technology in the semiconductor industry (Harris 1970). TI's management methods constitute the principal technique for integrating R&D with production and marketing. The company balances the short-run profit motive with the long-run objectives through its Objectives, Strategies, and Tactics (OST) system. The OST system entails two reporting modes: a strategic (long-run) and an operating (day-to-day) mode. Thus, the manager is accountable for both long-range growth and short-term profitability.

The OST system at TI calls for heavy emphasis on planning activities. This emphasis demands time and resources for strategic effort that produces conflicts with current operational priorities. As current and former TI employees have indicated, the management system there may be overdone and there is a danger that managers spend all their energies on planning and reporting (*Business Week* 18 September 1978).

The importance of organizational flexibility is also reflected in the fact that the semiconductor industry has remained relatively labor intensive. Skilled labor can adapt to changes in production techniques more readily than can capital. This fact is particularly true for capital in the form of specialized equipment. The experience of firms such as Philco that failed through investment in highly automated transistor facilities underscores this point. The relative labor intensity of the semiconductor industry compared to other industries can be seen in table 3-10.

In this section we have reviewed the relationship of organizational flexibility and corporate success in the digital integrated-circuit industry. The available evidence with respect to levels of organizational flexibility is indirect but points to the fact that an appropriate level of organizational flexibility is required for success generally and innovative success particularly. Firms that seemed to have the most trouble on the "too-inflexible" side of the flexibility optimum were semiconductor divisions of large con-

Table 3-10
Comparative Labor Intensity: U.S. Domestic Semiconductor Manufacturing versus All U.S. Manufacturing: 1967, 1972-1976
(*percent*)

	Payroll/Value of Output [a]					
	1967	1972	1973	1974	1975	1976
Semiconductors	46.7	33.5	31.6	32.1	38.2	30.4
All manufacturing [b]	21.9	21.0	20.0	18.1	18.3	17.8

Source: Census of Manufacturers, 1967 and 1972, and Annual Survey of Manufacturers 1973, 1974, 1975, and 1976, as compiled in U.S. Department of Commerce, *Report on the Semiconductor Industry* (Washington, D.C.: U.S. Government Printing Office, September 1979), p. 33.

[a] The value of output is equal to the value of U.S. domestic industry shipments adjusted for the inventory change over the year.

[b] The data are for operating manufacturing establishments and exclude central administrative offices and auxiliaries.

glomerates. Large, autonomously managed semiconductor firms also have had problems. Some industry experts suspect that TI's OST control system may be too inflexible. In constrast, Fairchild in the late 1960s appears to have been too loosely structured.

The style of management that appears to promote innovation in the semiconductor industry was succinctly summarized by an industry executive in one of our interviews as follows:

> A management style that permits geniuses to contribute is important. If you were to look at why GE and RCA have failed, it is because their organization was too disciplined and unable to respond quickly to true innovation. What is required is a balancing act—the organization must be loose and flexible, but not too loose. In the 1960s Fairchild was too loose. Today both Intel and Mostek are killing TI in new technology, in part because TI is too structured. Fairchild also seems to be lagging behind the leaders for the same reason.

The type of structure that is appropriate for a successful, innovative firm depends on a firm's size and strategy. An industry consensus seems to have emerged that communication channels must be strong between production, marketing, and R&D. In many firms this communication is accomplished with a structure composed of tightly knit product groups within a matrix organization. Whether this product-group structure serves to foster major as well as incremental innovations is unclear from the evidence we were able to collect.

Capital Availability, R&D Spending, and Risk Taking

The effects on innovation of the availability of capital and the level of R&D spending are difficult to separate analytically. Capital availability is a necessary prerequisite for R&D spending. Furthermore, innovation requires spending beyond the R&D stage, where capital availability is also a prerequisite. Since R&D spending is a risky investment, risk taking is also dependent upon capital availability. However, our results indicate that different firms undertake different levels of risk, suggesting that variables other than capital availability also influence risk taking. The remainder of this section discusses these issues.

Kamien and Schwartz (1978) present a useful model for evaluating the relation of capital availability to innovation. They define a cash constraint as a condition in which a firm's cash flow at a particular point in time is entirely consumed by spending on an innovation. They show that as long as an established firm is making money on its current product it can internally finance a new product that represents a routine advance. An order of magnitude for a routine advance is one that would no more than roughly double expected profits. However, very large innovations by established firms or efforts by new or marginally profitable firms would be constrained by cash availability.

These theoretical predictions can be supported by evidence from the integrated-circuit industry. First note that the industry is characterized by both routine incremental innovations and major innovations. The theoretical prediction is that most established firms will have little difficulty financing incremental innovations and in fact, the record shows large numbers of improvements by a diversity of firms (see the section titled The Innovative Record and chapter 4). A corollary of this prediction is that a certain level of relatively routine R&D will be undertaken by all established firms.

New firms will be able to engage in incremental product innovation if they can obtain sufficient external financing or if they can build initial profits through imitating established firms' products and then later introduce their own designs. Examples of both situations are found among new integrated-circuit firms. However, new firms will be able to make major innovations only if they have substantial outside financial backing, as was the case with Intel and Fairchild.

Major innovations, however, need not cost large absolute amounts of money. Perhaps the most important digital integrated-circuit innovation of 1970, for instance, was Intel's 1103, a 1,000-bit MOS dynamic random-access memory. This product was pivotally important because it was the first successful product in a series of MOS random-access memories that was targeted at computer applications, heretofore performed by magnetic-core memories. It successfully opened up a new demand for digital integrated circuits and became an industry standard. In 1970, when Intel introduced the 1103 its R&D spending was on the order of $1 million per year. In contrast, the rest of the semiconductor industry at that time was spending about $100 million on R&D. However, $1 million was a large sum for Intel at that time because the firm had just begun operations. Only because the company had backing was it able to carry through the innovation (Bylinsky 1973).

The data on R&D, patents, and major innovations shown in table 3-11 suggest that R&D spending is much more closely related to incremental than major innovation. The number of patents a firm obtains should be a relatively good measure of its output of incremental innovations. As seen in the table the number of patents appears to be much more closely correlated with cumulative R&D spending than is the number of major innovations. However, because major innovations are relatively rare, a larger sample would be necessary to make strong conclusions about the relation between R&D spending and major innovations.

The history of digital integrated circuits abounds with innovations that were not successful in the marketplace. Some of that history is summarized in chapter 4. The fact that so many innovations prove to be commercially unsuccessful indicates that spending resources to be an innovative leader is a very risky venture. Indeed, in a sense, the decision to spend for R&D is equivalent to a decision to take financial risks. Although the amount of in-

Table 3-11
Comparison of Cumulative R&D Expenditures, Semiconductor-Device Patent Applications, and Significant Innovations for Nine Firms: 1972-1976

Firm[a]	R&D Expenditures[b] (millions of dollars)	Patent Applications[c]	Major Innovations[d]
AMD	6	1	0
American Microsystems	24	10	0
Mostek	16	13	2
National Semiconductor	53	33	0
Intel	54	32	5
Fairchild	193	33	2
General Instrument	93	12	0
Motorola	455	147	0
Texas Instruments	280	108	2

[a]The last four firms are large diversified companies in which significant portions of their R&D expenditures will be allocated to projects outside their semiconductor operations.
[b]All R&D data are from Dataquest, except for General Instrument and Motorola. Their data come from company 10K and annual reports.
[c]Office of Technology Assessment, U.S. Patent and Trademark Office, device-related semiconductor patents (see table 3-4) for patents granted through 1978.
[d]From tables 3-1, 3-2, and 3-3 of this report.

Table 3-12
Responses to Questions Concerning Changes in Risk over Time

Direction of Change	Explanations Offered for Change	Number of Companies
Risk has decreased	Major areas of technology charted. Markets become more evolutionary and stable.	2
Risk has increased	Technology has become complex.	1
	Greater complexity, longer development effort.	1
	Cost of implementing innovation and necessary equipment has increased.	2

Source: CRA interviews with industry executives.

formation we have been able to obtain on risk is limited, we have found important insights. First, evidence exists, as hypothesized, that firms taking greater risk achieve greater innovative output. Second, although all the firms we interviewed use some form of quantitative techique incorporating risk in their innovative decisions, most executives stressed the highly subjective nature of risk assessment. Finally, we have empirical evidence of an important relationship between risk taking and capital availability that was overlooked in the original framework but is consistent with the conclusion that undercapitalized firms are not successful innovators.

The financial backing for an innovation may come from either a firm's existing resources in the case of an established firm that attempts to innovate or from venture capitalists in the case of a new firm. In either case, capital is necessary to finance salaries and equipment for R&D and start-up manufacturing and marketing.

The experience of two firms, AMD and National, illustrates the relation between capital availability and risk taking. Both firms pursued relatively low-risk strategies, emphasizing second sourcing when they were small and strapped for cash. As both companies grew and became more financially secure, they undertook riskier projects. National was a small $5 million company in 1967 when it was reorganized around a top-management team consisting of Charles Sporck and several other former Fairchild employees. In the early years the new management undertook only projects that would pay back in less than six months. Today payback periods of several years are acceptable. Larger size allows a longer view and greater risk taking than when the company was small.

The president of AMD, Jerry Sanders, uses a paradigm that summarizes this relation of risk taking to financial health. When AMD started, cash was short so they planted cash crops like lettuce that would mature in a few months—that is, they second sourced. Revenue from its second-source products enabled AMD to finance development of proprietary products. Starting the past few years and continuing into the next decade AMD intends to plant asparagus, which takes two years to mature but is more profitable and produces for a long period of time—that is, AMD will innovate and produce proprietary designs (interview with industry executive).

Other evidence on the relation between finance and risk comes from responses to our question about whether the risk associated with innovation had changed over time. Table 3-12 summarizes these responses. Opinion was divided among the interviewees about the direction of change of risk. Some said that risk had decreased as technology and demand became more stable. Others said that the risk had increased as the technology became more complex and the cost of innovation increased.[3] These responses are not inconsistant, of course. The probability of success could increase as technology and demand stabilized while the required R&D cost per innovation also increased. Thus, even though the relative risk (probability of failure) decreased, the absolute risk (probability of failure times cost) could increase. What is interesting is the linkage of risk and R&D cost by many of the respondents. This further underscores the link between capital availability and risk taking hypothesized earlier.

During our interviews with semiconductor executives we asked three basic questions concerning risk. First we inquired into the techniques used in evaluating risk and making innovative decisions. We found that much of the decision-making process is subjective, and we did not find enough similarity or rules of thumb to facilitate comparisons across firms. The second question involved comparing the interview company's level of risk with other companies. This question produced some useful responses, which will be discussed shortly. The third question concerned the changes in risk over time that were just discussed.

Table 3-13 summarizes responses to the following question: Compared with other manufacturers of digital integrated circuits, have the R&D projects that your company has undertaken been of equal risk, greater risk, or less risk than the R&D projects of other firms? Several executives explained that their portfolio of R&D projects included some very high-risk projects and some low-risk projects, but that on average their portfolio was comparable to other firms' in terms of risk. Other executives felt that on balance their firm took greater risk than other firms. Many of the interviewees expressed difficulty in making the comparison and therefore the response should be judged cautiously. In addition, no responses stated a level below average, which indicates the possibility of bias. Nevertheless, the results are interesting.

As shown in table 3-13, seven usable responses to the question were ob-

**Table 3-13
Level of Risk Taking, Number of Key Innovations, and Number of Patents**

Level of Risk Taking	Number of Companies	Number of Key Innovations 1969-1976	Number of Patent Applications[a] 1969-1976	Number of Patent Applications per Million Dollars of Sales[b]
Above average	3	10	223	.057
Average	4	3	668	.106

Source: CRA interviews with industry executives. Innovation data from tables 3-1 through 3-3. Patent data from table 3-4. Estimates of firm semiconductor sales data from Tilton (1971) and Dataquest.

[a] Patents granted through 1978 that were applied for in 1969 or later.

[b] The cumulative (1968-1975) semiconductor sales estimates for the firms in the two groups were added. These two totals were then divided into the patent-application totals to derive the number of applications per million dollars of sales.

tained. Three firms indicated a level of risk taking above average while four firms indicated an average level. The table also compares the number of key innovations, number of patent applications, and number of patent applications per million dollars of sales between 1969 and 1976. The above-average risk group has a higher level of key innovations but a lower level of patents, even when adjusted for firm size. This evidence suggests that above-average risk taking is more important for major innovation than it is for incremental innovation.

All companies interviewed during this study use some form of quantitative techniques in making innovation decisions. However, most companies stressed the highly subjective nature of some elements of the process. Techniques in use involve evaluation of technical risk, evaluation of market risk, assessment of resource availability, assessment of competition, and analysis of timing. One company is currently developing new numerical methods to evaluate technical and market risk. Past models abstracted from competitive response, but the new ones will allow for it. Another executive stated that his company placed relatively greater emphasis on evaluating market opportunity and risk than on technical risk. The primary market risk, in his view, was in possibly not doing something and getting beaten. In evaluating the cost of an innovation, this company computes a ratio based on the prospective returns to the use of skilled manpower. Since skilled labor such as design and production engineers is in short supply, this company prefers projects that use less-skilled time if returns are equal.

A third company described its decision process as follows. The company has divided its business into twenty strategic segments. A planning committee works on each segment assessing the market and the competition. Each committee puts together a model for its segment. Projects are ranked annually using the planning committee's results. Then top management decides what level of investment can be sustained and projects are chopped from the bottom of the list. A rule of thumb for this company is that the potential sales over the next five years must be twenty times the development cost in order to approve a project.

Finally, another company applies a common format involving payback analysis to all potential projects. Considerable evaluation effort unique to each project is involved, however. This company's executive stated that due to abundant opportunities few projects with less than a 50-percent probability of technical success were undertaken. Frequently the probability of technical success was in the 70-percent range. This company also prefers to avoid technical risk and market risk in the same project. Their approach is to use a known technology for new markets; to attack a known market with a new technology, however, is acceptable and frequently necessary. This rule of avoiding simultaneous technical and market risk was not found in our discussions with other companies.

The subjective nature of innovation decisions is particularly apparent with respect to technical risk and timing. One of the leading researchers and executives in the industry stated that he has difficulty with a quantitative assessment of technical risk. Another company's executive said that the most difficult and least quantifiable part of risk evaluation is the proper timing of innovation. Timing is the area most frequently misjudged and the one most damaging to be wrong about. A product introduced before the time that it can be produced in the quantity and at the cost that customers want will fail. New product design started after competitors have started theirs can also fail if the first product of a type becomes standard. Other executives described risk evaluation and innovation decision making, particularly for major projects, as "good judgement biased by emotion," "gut-feel," and "seat-of-the-pants" analysis (interviews with industry executives).

The interviews we conducted with key industry executives confirmed our hypothesis that capital availability is a necessary prerequisite to the successful undertaking of R&D spending to develop risky innovations. We find in chapter 4 that many innovations are not commercially successful. Combining this insight with the insights obtained in the interviews with key executives has led us to the conclusions we have developed, namely:

If profits are adequate, incremental innovation can be financed out of internally generated funds. Major innovations may be constrained by cash-flow availability, however, particularly for small firms.

Major innovations do not necessarily require large absolute amounts of R&D spending. Much R&D spending is for incremental innovation.

Most semiconductor executives rely heavily on subjective criteria in judging the level of risk of a project. The proper technical concept and the proper timing for its introduction were noted as being important subjective standards by which to evaluate risk.

The number of major innovations is positively related to the executives' assessment of their company's level of risk taking compared to the industry average.

Other Characteristics that May Facilitate Innovation

The following subsections discuss four issues that were hypothesized to be important for innovation in the overall study framework but do not appear to be critical in the case of digital integrated circuits. The issues are patents, basic research, technical expertise of the marketing staff, and financial incentives within the firm. Although these issues do not appear to be critical to explain interfirm differences in innovation, they may facilitate overall industry innovation.

Patents. The overwhelming view among industry executives is that patents provide little incentive for innovation in the semiconductor industry. Two main reasons for the failure of patents to provide an incentive exist. One reason is that semiconductor patents are so numerous that firms have to cross license each other. Without cross licensing any of several firms could block others from producing, and wasteful legal confrontations would result. Second, the time required to obtain a patent is long in relation to the speed with which semiconductor technology changes. As one executive put it, "the payoff comes from charging ahead with the innovation. Patents may provide some protection later." On a related point, one executive noted that disclosure was not a disincentive to patenting. Patents are disclosed only when issued, not while pending. Since the technology changes rapidly and information spreads so fast in the semiconductor industry, it is usually possible to tie up an application long enough in the patent office to avoid any disclosure problem.

Several other factors also contribute to the low patent incentive. For example, process and production equipment innovations are frequently more valuable as trade secrets. Finally, one firm cited Justice Department opposition to the use of patents to obtain a dominant market position. As a result royalties have been relatively low. Firms have felt they risked antitrust litigation if they charged high royalties or otherwise used patents to build a market position.

Despite the failure of patents to provide an incentive for innovation, most firms obtain patents for defensive purposes. Patents are viewed as a bargaining chip in cross-licensing negotiations and as a means to minimize royalty payments. The larger firms generally patent most inventions, but many cited the cost of patenting as a deterrent. For instance, RCA files patents liberally. But maintaining patents can become expensive, so RCA prunes its patent portfolio periodically. Other large semiconductor manufacturers such as Intel, Motorola, National, and TI also expressed high propensities to patent. Only one of the smaller firms, AMD, stated that not all inventions were patented. This was attributed to a lack of resources, both in people and time (interviews with industry executives).

The propensity to patent also appears to have changed over time. The following comments by one firm's executives concerning this change are insightful. Over the last eight to ten years this firm has become much more selective and purposeful in applying for patents. In the 1950s and 1960s it applied for patents on almost every discovery. Two reasons for this change were suggested. First, process and equipment innovations are frequently more valuable as trade secrets. Second, the cost of obtaining a patent can be prohibitive. This second reason is particularly true for the broad international coverage new required due to the emergence of major competitors and technology practice in many other countries.

According to these executives, in the 1950s and 1960s most patent negotiations were entirely among U.S. firms. As other countries caught up, however, U.S. companies found it necessary to build strong patent positions abroad. This factor has greatly increased the cost of patent filings because filings are necessary in about ten countries. Thus, U.S. firms tend to pick and choose what they patent.

Also according to these executives, the number of patents as an indicator of innovation among countries is currently biased by this selectivity of U.S. firms. In addition, Japanese firms are aggressive in filing for U.S. patents. This aggressiveness is probably due to the high royalties the Japanese have faced. Japanese firms have been paying roughly 10 percent of their semiconductor sales in royalties, primarily to Fairchild, RCA, TI, and Western Electric.

In the 1980s this Japanese aggressiveness and U.S. reluctance to patent could be a major problem. Assessing the long-term impact of a patent is very difficult. The executives stated that numerous examples exist of patents that did not look promising initially, but later the company was thankful it had filed. The interviewees would not be surprised if the Japanese "shotgun approach" resulted in some important patents. A reverse flow of royalties to Japan could well develop in the 1980s. On a related point, another executive noted that his company aggressively asserts its patents in Japan to keep Japanese companies from restricting his company with patents in the United States (interviews with industry executives).

Some of the firms that obtained strong patent positions early in the semiconductor-industry history earned large royalty income and other benefits such as access to foreign markets and technology. Although difficult to prove conclusively, these benefits may have facilitated subsequent innovation. Western Electric held many important patents in the 1950s and readily offered licenses to interested firms. After 1953 it charged a maximum royalty rate of 2 percent of semiconductor sales (Tilton 1971, p. 74). In the 1960s Fairchild actively marketed licenses for its important planar process and integrated-circuit patents. Though it often provided considerable technical assistance to licensees, it charged royalty rates of up to 6 percent of sales (Tilton 1971, p. 77; Finan 1975, p. 47). Fairchild earned a healthy royalty income; according to Finan, it earned $6 million in royalty income in 1972.

TI also holds important patents for the integrated circuit, but it has been less aggressive in marketing licenses. TI, however, used its patent position to obtain a subsidiary in Japan in 1968. In exchange for licensing integrated-circuit patents to Nippon Electric, Hitachi, Mitsubishi, Toshiba, and Sony, and agreeing to limit its production to 10 percent of total Japanese production, TI obtained the consent of the Japanese government to hold a 50 percent interest in a joint venture with Sony (Tilton 1971, pp. 146-147). Subsequently, Sony sold its interest to TI, making TI the only U.S.-based company with a wholly owned manufacturing facility in Japan.

Basic Research. The National Science Foundation defines *basic research* as research without immediate commercial application: "the cost of research projects which represent original investigation for the advancement of scientific knowledge and which do not have specific commercial objectives, although they may be in fields of present or potential interest to the reporting company." Under that definition, the firms we interviewed conducted relatively small amounts of basic research. One interviewee said that some of the R&D conducted by his and other firms would undoubtedly fit a definition of basic research but that the proportion would be small. A second executive estimated that 5 to 10 percent of all companies' R&D was basic research. A third executive said that essentially none of his firm's R&D would fit the definition of basic research, yet 10 to 15 percent of its R&D was done centrally because it was too venturesome to be done in product divisions. "Fundamental" research conducted centrally was noted by a number of interviewees. For those who gave figures, the proportion of R&D done centrally ranged from 10 percent to one-third of total R&D. The firm reporting the highest proportion of research done centrally described it as "fundamental but with an ultimate goal in mind."

According to Finan (1975, p. 38) in 1972 $36 million was spent by private firms on semiconductor basic research. Bell Laboratories spent $15 million; IBM, $13 million; RCA, $4 million; TI, $2 million; and others, $2 million.

Innovative Behavior by Manufacturers

Some companies that we interviewed relied primarily on other companies for basic-research advances, while others relied heavily on universities. For example, several executives said that other companies such as Western Electric, RCA, IBM, and General Electric were the main source of basic research advances. For some of the companies we interviewed, universities were the second most important source of basic research, but for others universities ranked third behind the interviewed company itself. However, two companies rated universities as the primary source of basic research. One of these two estimated that 70 percent of the basic research that flows to his company comes from universities, while 30 percent comes from other companies. The other executive said that today universities were the main source of basic research, with his company second, and other companies third. In the past, the order was his company first, other companies second, and universities third.

In the 1950s and early 1960s, many of the key semiconductor innovations were contributed by firms such as Western Electric (Bell Laboratories), General Electric, RCA, and IBM that conducted substantial basic research. In the early 1960s Fairchild also conducted a substantial amount of basic research and was responsible for many key innovations. As time passed, however, the role of these firms and basic research declined. This trend can be seen in tables 3-1 and 3-2. This decline reflects several phenomena. As a branch of technology develops, much of the required basic research may accumulate and not need to be repeated. Also, as the commercial importance of a field becomes apparent, much research that advances science as fundamentally as previous (basic) research may no longer fit the rigorous definition of basic research. As opportunities become apparent, the research falls under the definition of applied R&D.

Interestingly, although RCA and Fairchild are credited with a lot of work in the early 1960s that led to MOS integrated circuits, they were unable or unwilling to introduce the technology to the marketplace. Firms such as GMe, AMI, and General Instrument, which conducted less basic research, were the first to exploit MOS technology commercially (Kraus 1973, pp. 51-54). However, the reader should note that firms like GMe were founded with a nucleus of Fairchild's MOS people. The loss of these people may have delayed Fairchild's push into the MOS market. As is explained in chapter 4, the MOS integrated-circuit technology did not really make any major market penetration in any event until the very late 1960s.

Judging the value of basic research by the value of innovations produced in the firm doing the research may, however, understate the significance of basic research, if basic research is to some degree a public good (Arrow 1962). Firms in the industry that do not conduct basic research may be free riders of other firms' basic research. Arrow points out that if the knowledge produced in research cannot be easily assigned a "property right" that can be bought and sold, the research result is readily available to all companies and thus is a "public good." The great amount of interfirm personnel mobility in the semiconductor industry means that hoarding knowledge is difficult and research results are therefore in Arrow's sense a public good. Judging basic research by its value to its funder also understates the importance of basic research in that fundamental research often leads to innovations relevant to other industries.

Technical Orientation of Marketing Staff. In most industries customer contact is important for successful innovation in at least two ways. First, information about user needs is necessary to design new products. Second, once a new product exists, its capabilities must be communicated as part of the selling effort. In both cases, technical proficiency of a firm's marketing staff is necessary. One might therefore expect some firms to attempt to gain an edge in innovation through recruiting a highly technical marketing staff.

Contrary to expectations, however, little difference exists among semiconductor firms in the technical orientation and background of marketing personnel. The function of understanding user needs appears to be largely accomplished through the management techniques discussed earlier, such as the division of staff into product groups in the matrix organization or TI's OST system, that integrate R&D, production, and marketing. The function of communicating the capabilities of new products is generally accomplished by having technical specialists available to be called in by the front-line salespeople when needed (interviews with industry executives).

Nearly all the executives we interviewed rated their company's marketing staff as average for the industry. The main exception is Intel, whose marketing staff is hired to be more technical than average, primarily because selling microprocessors involves selling computer-systems concepts (interviews with industry executives). However, other companies also noted that greater technical expertise than average is required to sell their microprocessors. The difference between Intel and other firms seems to be that most other companies have a greater proportion of older products, which require less technical expertise to sell, than does Intel.

The views of one company's executives with respect to the selling function are instructive. This company recruits primarily engineers for sales positions. The company provides its own training to supplement the salesperson's education. However, the sales force is not designed to be highly technical. The company wants salespeople to be effective in closing sales, managing customers, and in general using their time effectively. For example, if a salesperson spends several hours in a customer's back room talking to designers, that activity is not viewed as effective use of time. The company will thus get that person out of sales and into some other part of the company. The company does, however, use the technical-specialist function where technical support is needed by customers, such as in the microprocessor area.

Financial Incentives within the Firm. The financial rewards realized by individuals in the semiconductor industry have not been overwhelming. In a few cases individuals have done well, mainly through forming new companies. However, on balance salaries and incentive compensation are in line with averages for other manufacturing industries (see appendix B). On reflection this fact is perhaps not too surprising. Although the pace of technological change has been rapid, so has the decline in prices resulting from intense competition. Few economic rents are left for either the owners of capital or for individuals.

This does not mean, of course, that the lure of getting rich was not a major motivating factor in the industry. In fact, many people we have talked to believe that it was. Although average salaries are comparable with other industries, this comparison can be deceptive due to differences in age structure. Managers in the semiconductor industry are young, bright, and self-confident. This age factor was more important in the past than it is today, but young, talented managers view the semiconductor industry as a place where they can obtain rapid position and salary advancement. Thus, even though average salaries are not unusual per management level, the ease of obtaining higher levels is unusual because of the industry's rapid growth. Thus, expected rate of salary growth rather than current salary level can be a key incentive to recruiting the brightest young engineers and managers. But, the failure of many to strike it rich does explain the fact that the dropout rate of engineers and managers from the industry is said to be very high.

A major change in the tax treatment of stock options occurred in 1969. (See chapter 6 for a more lengthy discussion.) Many observers believe that this change greatly affected individual motivation. One executive pointed out that relatively few people became really rich in the industry either before or after the tax change. However, before the change nearly everyone was trying to get rich, while afterward far fewer people were. The result is less diversity in new ideas and innovative effort (interviews with industry executives).

The financial incentives to start a new firm are closely tied to tax laws and the availability of venture capital. The effect of these issues on the rate of formation of new semiconductor companies is discussed in chapters 2 and 6.

Appendix B disscusses incentives within firms in the semiconductor industry. It presents data on salaries and stock holdings of top industry executives and average salaries of white-collar employees. Nearly all firms provide incentives for key employees based on corporate performance. However, except for Intel, the rewards achieved by individuals are similar across companies and comparable to other industries.

The Intel example raises the question of how important financial incentives within a firm are for innovation. Causality clearly flows both ways. Incentives encourage innovation, but the profits accompanying innovative success also allow employees to exercise incentives such as stock options. Since other firms provide similiar incentives, much of the explanation for Intel's success must be attributed to other factors.

Financial incentives within the firm, then, may facilitate innovation but do not appear to be one of the most important characteristics of innovative firms. The fact that few people achieved outstanding financial success suggests that while the potential rewards may need to be high, the odds of achieving them do not need to be high in order to motivate people. As mentioned above, expected growth rate of salary can be quite important, although we have no direct evidence related to this hypothesis.

Summary. In this subsection we have examined the role that patents, basic research, technical expertise of sales staff, and financial incentives play in digital integrated-circuit innovation. While we initially hypothesized that these factors would be important with respect to innovation, we have found that they do not play a major role.

Patents seem to be more a source of royalty revenue than an incentive to innovation. Firms weigh the cost of filing for and maintaining patents against the possible royalty gains and against the risk of domestic and foreign competitors' patent royalty demands. Some industry executives are concerned about the current aggressiveness of the Japanese taking out U.S. patents. The main feature that negates patents as a stimulus to innovation is that the length of time required to obtain a patent is long in relation to the speed with which semiconductor technology changes.

The importance of basic research in promoting innovation appears to have declined since the early stage of the industry. However, merchant semiconductor firms do some basic research. In addition, larger firms not active in merchant integrated-circuit sales, such as IBM, Western Electric, and General Electric, do considerable semiconductor basic research today. These firms and RCA, which is active in merchant sales, were cited as the primary source of basic-research results for the industry. In addition, universities are an important source for many firms.

The technical expertise of salespeople was not considered important in promoting innovation. The probable reasons for this conclusion are that the firm's marketing organization can call on technical specialists as needed. A minimal level of sales-staff technical competence is desirable, but outstanding technical credentials are not required.

Financial incentives as measured by data on average salaries and number of individuals who have achieved substantial wealth suggest that certain monetary rewards are not one of the prime explanations for innovation. Rather, the slim chance of becoming wealthy combined with adequate salary levels may be the key. In addition, the expected rate of growth of salary rather than the salary per level of management may be very important. The people we interviewed generally believed that a major disincentive to extra effort and risk taking was the change in the tax treatment of stock options that occurred in 1969.

Statistical Testing of the Effect of Firm Characteristics on Innovation

This section presents a straightforward statistical model based on the previous section's analysis of innovative behavior. Essentially, we test for the effect of three of the most important firm characteristics on two alternative measures of innovative output: key innovations and patents, the latter a measure of incremental innovation. The three characteristics are: (1) top-management commitment to innovation, measured by the portion of top management's time devoted to innovation-related matters; (2) degree of risk taking, measured by executives' assessment of their firm's risk taking compared to the rest of the industry; and (3) R&D spending and capital availability, measured by the firm's total semiconductor sales.

The main characteristic that is not included in this analysis is the organizational-flexibility/management-control balance. The reason for excluding this variable is that we are simply unable to quantify organizational differences among the firms for which we could obtain data on the other variables. These firms have all survived in the semiconductor industry and to some extent passed the organizational and management test that screens out firms. Of the firms for which we obtained data, RCA is the one that differs most from the others in terms of organization and management. However, we felt it necessary to exclude RCA from the analysis for reasons elaborated in the discussion of innovation data in this section and in chapter 4.

After briefly describing the data, we present the econometric results. The main limitation of the data is the relatively small number of observations; essentially only eleven observations are available. Therefore, difficultly in obtaining results that are statistically significant can be expected. Nevertheless, the results suggest that top-management commitment and degree of risk taking have their strongest effect on key innovations, while R&D spending and capital availability most strongly affect incremental innovations.

The Data

The key-innovation dependent variable is constructed from the data in tables 3-1 through 3-3 and is the total number of innovations a firm introduced during a given time period. To increase the number of observations we considered two distinct periods of innovations: 1960-1965 and 1968-1975. Three of the companies for which we could obtain data existed in the earlier period and eight in the latter period, for a total of eleven observations. Although we were able to obtain data on independent variables for one additional company (RCA), this observation was eliminated because its innovative activity is very difficult to categorize. This problem is discussed further in chapter 4.

The data in tables 3-1 through 3-3 support the division into two periods. As can be seen from those tables, few innovations occurred in 1966 and 1967. Further support is found in the information presented in chapter 4. The early 1960s were a time of innovation in bipolar circuits, while the late 1960s and early 1970s saw much activity in MOS technology.

Two alternative methods of counting the key innovations are used. One is to simply add up the innovations in tables 3-1 through 3-3 for each firm. This measure is denoted "INN1" in the presentation of results. The second method attempts to eliminate double counting by: (1) counting as .5 any innovation attributed to more than one firm; and (2) counting as 1.0 (instead of 2.0) any pair of product-family and device-structure (or process) innovations that are closely related. This measure is denoted "INN2." The difference between the two measures mainly affects firms with several innovations. For instance, in the 1960-1965 period, Fairchild has seven under the first measure and four under the second.

The other measure of innovative output that we use is the number of patents obtained by a firm. Although we found in the previous section that patents were of relatively minor importance in the semiconductor industry, our interviews showed that most firms obtain patents on their inventions. Therefore, the number of patents provides a useful record of the number of advances. Since far more patents than key innovations exist, patents are more a measure of incremental innovation than the previous measure.

A proxy for patent applications for the 1960-1965 period is obtained from Tilton (1971). Since we wish to relate innovative activity as closely as possible to the years covered, we assume a lag of three years between patent application and issue and therefore use patents granted from 1963-1968.[4] For the period 1968-1975 we use the patent-application data for semiconductor devices from table 3-4. We believe that this category provides the closest correspondence to the data from the earlier source.

The top-management time allocation and degree of risk-taking variables were discussed in the previous section (see tables 3-8, 3-9, and 3-13

Innovative Behavior by Manufacturers

and accompanying text). The data we collected on these variables were assumed to apply to the 1968-1975 period, since the executives interviewed were at the companies during that time. To extend the measures to the 1960-1965 period, two assumptions were necessary. Where management was relatively stable, such as at TI, the same data were assumed to apply to the earlier period. Where the executives we interviewed had changed firms and had been the top people at other firms in the earlier period, we applied their data to the other firms in the early period.

Total semiconductor sales is used as a proxy for R&D spending for several reasons. First, differences in the definitions that firms use in presenting R&D data make it difficult to compare R&D figures across firms and over time. Few semiconductor firms average much more than 10 percent of sales on R&D or much less than 5 percent of sales over any length of time. Furthermore, when an attempt is made to eliminate definitional differences, many of the firms' figures converge, while some remain a few percentage points higher. A much greater effect on the variation in firms' R&D spending occurs due to the variation in firm size. Therefore, total sales is a reasonably good proxy for total R&D spending. A second reason for using sales is that R&D data are not available for the early 1960s, while sales estimates are.

Statistical Results

Table 3-14 presents regression results showing the effect of the top-management involvement (TMI), RISK, and SALES variables on the alternative innovation measures. As discussed in an earlier section, the top-management characteristics and degree of risk taking are closely related. The statistical correlation between the TMI and RISK variables is .564. In contrast, the correlation of sales with TMI is −.084 and with RISK is .098. Therefore, results are shown for equations that exclude one or the other of these variables in order to demonstrate the effect of collinearity.

In the key-innovation regressions, the coefficients of TMI and RISK are each statistically significant when the other variable is excluded.[5] When both variables are included, the statistical significance drops due to the collinearity. However, the SALES variable is not statistically significance in any of the key innovation regressions. A different picture emerges from the patent regressions, where the sales coefficient approaches acceptable levels of statistical significance. However, the TMI and RISK coefficients are very weak.

Although these results are limited by the small number of observations, they are instructive. A high degree of top-management involvement and above-average risk taking promote major innovations. Capital availability

**Table 3-14
Statistical Effect of Firm Characteristics on Key Innovations and Patents**

Dependent Variable	Regression Coefficients[a] (t-statistics in parentheses)				
	Constant	TMI	RISK	SALES[b]	R^2
INN1	−0.94	0.09	2.28	0.10	.575
	(−0.54)	(1.06)	(1.68)	(0.18)	
INN2	−0.06	0.04	1.31	−0.08	.500
	(−0.06)	(0.80)	(1.53)	(−0.24)	
Patents	36.50	−0.20	18.00	43.72	.349
	(0.50)	(−0.06)	(0.32)	(1.87)	
INN1	−1.79	0.17	—	0.19	.402
	(−0.97)	(2.32)		(0.31)	
INN2	−0.55	0.09	—	0.03	.331
	(−0.48)	(1.98)		(−0.08)	
Patents	29.76	0.44	—	44.43	.339
	(0.45)	(0.17)		(2.03)	
INN1	0.64	—	3.13	0.03	.506
	(0.70)		(2.86)	(0.06)	
INN2	0.69	—	1.72	−0.12	.454
	(1.25)		(2.56)	(−0.34)	
Patents	32.86	—	16.06	43.88	.348
	(0.93)		(0.38)	(2.02)	

Source: Calculations by CRA, January 1980.
[a]All regressions with eleven observations.
[b]Sales in billions of dollars.

and R&D spending, both proxied by total firm semiconductor sales, have little effect on major innovations, but they have a much stronger effect on incremental innovations.

Conclusions and Policy Implications

This chapter analyzed a number of firm characteristics that were hypothesized to be closely related to innovation. We found that in the case of digital integrated circuits some characteristics were more important than others. In particular, top-management background and commitment to technology, the degree of risk taking, the proper balance between organizational flexibility and management control, capital availability, and R&D spending were found to be associated with innovation.

The background and talents of top management give particularly important insight into innovative behavior and competive strategies in general. Some top executives are simply more insightful or interested in technology than others and the behavior of their firms reflects this facet. In addition, we found that the amount of time allocated to innovation-related matters by top management was a good indicator of innovative behavior.

In examining risk taking we found that semiconductor firms have a portfolio of R&D projects with different levels of risk. Many executives said that the overall level of risk their firm undertook was similar to other firms. However, some executives felt that their overall level of risk was higher than the industry average. This higher level of risk taking was also associated with a higher degree of top-management involvement in innovation.

The organizational form that has had the most difficultly with innovation in the semiconductor industry has been the nonautonomous division of a large diversified firm. However, autonomous semiconductor operations have also had problems in the past. Most firms have learned from earlier firms' problems and adopted organizations characterized by tightly knit product groups with R&D decentralized and conducted within the product groups. Only the most fundamental and venturesome R&D is typically conducted centrally today.

If profits are adequate incremental innovation can be financed out of internally generated funds. Many small firms, however, may be constrained by capital availability with respect to introducing major innovations. We also found that major innovations do not necessarily require large absolute amounts of R&D spending and that much R&D spending in the semiconductor industry goes for incremental innovation.

In addition to demonstrating the importance of these characteristics compared to other less-crucial characteristics, the chapter also showed that the important characteristics have different effects on different types of innovation. Top-management involvement and a high overall level of risk taking were found to be closely associated with major innovations. In contrast, capital availability and R&D spending were found to be more closely associated with incremental innovation.

The results found here have important implications for federal government policy. Government policy can have a direct effect on incremental innovation due to the importance of capital availability and R&D spending. These variables are within direct influence of the government through its tax policies and interest rates.

However, the ability of government policy to influence major innovations is much more indirect because the important variables influencing major innovations are top-management background and commitment and a high degree of risk taking. The major effect must come through policies that encourage people with the right talents and inclination to reach the top spots in firms that face the opportunities and have the organization to achieve major innovations. The indirect effects of policy variables are hard to identify, but several can be mentioned: capital gains or stock-option incentives, minimal legal restrictions on personnel mobility, and low patent barriers. Note that an important mechanism for these policies is through encouraging technological entrepreneurs to spin off and start at the top of a new firm.

One means by which government policy can induce greater risk taking is to increase the rewards successful entrepreneurs can expect. Several executives of semiconductor firms whom we interviewed cited the importance of the entrepreneur's assessment of risks and rewards in making decisions related to innovation. The greater the rewards that can be expected if successful, the more inclined the entrepreneur is to undertake the risks necessary to achieve success in the market.

The government can also use its policy instruments to reduce the risks entrepreneurs confront. As we discuss in chapters 1 and 6, in the early years the government funded much R&D activity related to semiconductor technology and also offered lucrative procurement contracts to firms able to develop new semiconductor devices. Government support of R&D activity reduced the technical risk faced by the firm, while procurement contracts diminished the market risk.

Several of the characteristics that were found to be less crucial for innovation in the semiconductor industry also provide important policy implications. The history of patents in the semiconductor industry suggests that an industry can be innovative without strong patent incentives. More work is needed, however, to know how unique this observation is to the particular industry studied. For instance, the short product life cycle in semiconductors may make patents with seventeen-year lives and one-or-two year application lags irrelevant. In other industries with longer product cycles, patent incentives may be more important for innovation.

Basic research appears to be important for innovation in the birth and childhood of an industry. However, as the industry matures, firms that do relatively little basic research can produce major innovations. Finally, the comparability of semiconductor-industry salaries to those in other industries shows that people will spend considerable effort on innovation without any guarantee of unusual financial rewards. A few people captured some portion of the financial rewards from innovation, but most people did not because competition passed the rewards on to customers. The lesson for government policy is that policies such as capital-gains taxation need to permit a reasonable chance of striking it rich; competition will take care of the possibility that too much of the benefits of innovation end up in the pockets of the innovators.

Notes

1. Monolithic integrated circuits contain all components within the semiconductor chip. Hybrid circuits have one or more chips connected together within a device package.

2. TI's Jack Kilby invented a hand-wired integrated-circuit slightly earlier and the rival inventions led to a lengthy patent dispute. However, the planar integrated-circuit was more practical commercially.

3. There was no strong relationship between the responses to this question and the relative levels of risk taking discussed earlier. For example, of the two companies that said risk had decreased over time, one appeared in the previous tabulation in the above-average risk category while the other was in the average-risk-class.

4. Although the OTAF Patent and Trademark Office currently reports an average lag of nineteen months, this lag was significantly longer before the late 1960s.

5. The t-statistics shown in parentheses in table 3-14 allow tests of statistical significance. A t-statistic of approximately 2.3, for instance, indicates that the estimated coefficient is significantly different from zero at the 95-percent confidence level (two-tailed test) when there are 8 degrees of freedom.

4 Strategies and Competition Among Integrated-Circuit Firms

Accompanying the rapid pace of innovation in the integrated-circuit industry has been an intense competitiveness among firms. With each new round of product development, several firms typically vie to have their product design accepted by customers as the industry standard. As soon as a clear favorite emerges, a number of additional firms enter the race and adopt the standard design. Through price competition, the introduction of improvements on the standard design, and the use of marketing and distribution strengths, these firms strive for a share of sales in the new-product area.

This chapter analyzes the relationships among strategies and the interactions among different firms by conducting case studies in five digital integrated-circuit-product areas: bipolar logic, custom LSI, memories, microprocessors, and consumer product circuits. The first section discusses the key elements of strategy. The second section presents the case studies, and the third section presents the conclusions drawn from the case studies.

We have three major conclusions. First, strategic differences among firms have led to vigorous competition in the semiconductor industry. Second, in each of the technology races examined, the nature of the competitive process changed over time in ways consistent with the product lifecycle model. Finally, evidence in this chapter and chapter 3 suggests that firms can be classified into "strategic groups," the members of which share common elements of strategy. These strategic groups form the basis for analyzing performance in chapter 5.

Key Elements of Strategy

This section briefly outlines the essential elements of strategy that appear to be important in digital integrated circuits. The discussion is divided into four parts: product strategies, pricing, marketing and distribution, and vertical integration.

Note: Marvin Lieberman wrote the sections entitled "The Memory Race," "Strategic Diversity," and "Competition and the Product Cycle in Integrated Circuits." David Howe wrote the sections entitled "Microprocessors" and "Consumer Products."

Product Strategies

Three important dimensions of product strategies exist in the semiconductor industry: product design, reliability, and breadth of product line. The first dimension is related directly to innovation: the firm may choose to offer a product design of its own development or use another firm's design. The latter choice is commonly referred to as *second sourcing* in the industry; in other words, the firm is a second source of supply for the original firm's design. Two terms are applicable to the situation in which a firm develops its own design. The first is a *proprietary* design for a standard application. The second is a *custom* design, or a product design that is made specifically for a particular customer and not offered to other customers. Occasionally modified versions of custom circuits are offered to the general market after the initial design funding-contract terms have been fulfilled.

Most of the product advances that fit the proprietary (or custom) label could be called incremental innovations. These advances improve certain product features but do not necessarily represent a major technological breakthrough. However, as discussed in chapters 3 and 5 and evidenced in the following case studies, the cumulative total of these incremental advances may outweigh the improvements in product features and cost arising from major innovations.

The second important dimension of product strategy is reliability. The failure rate of a semiconductor device depends on such things as wafer-production-area cleanliness, bonding quality, and package mechnical integrity (see chapter 2 for a discussion of integrated-circuit production). The buyer has two reliability concerns: the device functions electrically but poorly, and the device fails completely through a mechanical breakdown (for example, a ruptured package seal). Reliability is usually measured as the probability of failure in a certain number of operating hours or as a mean time between failures. Extra testing does not improve the reliability of a single unit, but rather of a population of parts by culling out electrically or mechanically weak units.

Greater reliability can be purchased at the expense of greater testing and more elaborate process control. As with most activities, there are diminishing returns to these investments, and the question for manufacturers is: How close to 100 percent reliability is optimal? The answer will differ for firms pursuing different product-design strategies and different types of customers. For instance, greater reliability will help a second source sell its product but much more so in the military than in consumer markets. Although data on reliability at the firm level are sensitive and difficult to obtain, some information suggests a greater emphasis on reliability by some firms than others. AMD entered the industry in the late 1960s through second sourcing; one of its promotional features was to build nonmilitary products to military specifications, which was not the accepted industry practice at the time (interview with industry executives). In addition, Japanese firms' emphasis on reliability has given them a valuable reputation among U.S. buyers and is forcing U.S. semiconductor firms to reevaluate their own priorities between price and reliability (*Business Week*, 12 March 1979 and 3 December 1979, p. 85).

Breadth of product line is an important dimension of product strategy for several reasons. Product design and management resources may either be spread over a large number of products or concentrated on a small number. Similarly, demand uncertainty may either be spread over a number of products or greater risk taken by concentrating on fewer products. Breadth of product line also interacts with other strategic elements. Some marketing and distribution economies arise from having a broad line of products. In addition, firms that attempt to supply a full line of integrated-circuit devices will necessarily engage in some second sourcing. The industry is nowhere near the polar situation in which a single firm originates the propriety design for every single product.

Pricing

Pricing can be used in conjunction with other strategies in a number of ways. One empirical issue in whether well-defined and stable relationships exist between pricing and other strategies. A related issue is whether pricing may be used to substitute for other strategies such as innovation.

A firm's pricing behavior may be either aggressive (with relatively low profit margins). or passive (with higher profit margins). A firm introducing its own new proprietary design may adopt either of these pricing modes. The more aggressively it prices, the faster the new product will become accepted, and the more difficult it will be for other firms to enter. Passive pricing by the proprietary firm will allow second sources to gain market share but will give the proprietary firm greater profits in the interim.

Firms that second source another firm's design must generally be more aggressive pricers to gain sales. The main exception occurs if demand is so small that only one second source enters; in that case the second source may be more passive and higer priced than the primary supplier.

Older products would also seem to provide firms with an option between pricing aggressively to keep the product viable and pricing passively to reap greater profits. However, experience seems to have taught semiconductor firms the futility of maintaining aggressive pricing on an older product in the face of rapid technological change. Therefore, as other firms withdraw so as to free resources for other products, the remaining firms seem to opt for the passive strategy — maintaining prices or allowing them to rise. Motorola, for instance, has found adherence to older products longer than other firms particularly profitable (interview with industry executive).

Marketing and Distribution

The single most important aspect of semiconductor marketing and distribution is probably delivery. Timely delivery is frequently worth more to customers than new-product features or an edge in price or reliability, because most users of integrated circuits insert them into equipment in assembly-line factory operations. Integrated circuits are intermediate goods that must be delivered on time to avoid costly assembly-line shutdowns. Since delivery problems usually result from production-yield problems, delivery is clearly affected by variables throughout the firm. Firms that can control their operations carefully enough to assure good delivery gain an advantage over firms that cannot.

Another important aspect of marketing and distribution is the advantages of large size and breadth of product line. Larger firms can afford more extensive marketing and distribution networks. A salesperson serving a customer's needs for one product may discover or respond to a need in a different area. The existence of an established marketing and distribution network also makes it easier to introduce a new product. Potential buyers are known and quickly informed by a firm with a large, well-developed marketing organization.

Vertical Integration

Vertical integration between integrated-circuit manufacturing and downstream (or forward) activities such as manufacturing and selling electronic equipment or consumer products provides the main instance of vertical integration in the semiconductor industry. Most large semiconductor firms have attempted at one time or another to integrate downstream with parts of their product lines. In addition, a large number of equipment manufacturers have captive semiconductor production, own small merchant semiconductor firms, or own a portion of larger merchant firms. In contrast, upstream (or backward) integration into silicon production and process equipment in less common, with TI being the main example.

Captive production is an important portion of total integrated-circuit production. The two largest captive producers are IBM and Western Electric. At one time IBM was the world's largest semiconductor producer. A large number of other computer manufacturers, however, also have some semiconductor facilities. In addition, other equipment manufacturers such as Hewlett-Packard and Eastman Kodak also have captive facilities. Since much of the reason for captive production is to gain an advantage in equipment markets, analyzing these captive firms' semiconductor strategies is difficult, other than to note the existence of captive production as a strategy. However, the captive producers can be important in innovation, as seen by referring to tables 3-1 through 3-6.

In addition to expanding directly into equipment manufacturing, many smaller integrated-circuit firms have been acquired in whole or in part by large equipment firms, many of which are based in Japan or Europe. For instance, Nippon Electric Corporation owns Electronic Arrays, Philips owns Signetics, and Siemens owns 20 percent of AMD. U.S. equipment firms such as Honeywell and even aircraft firms such as Boeing and United Technologies have also acquired semiconductor firms.

Several potential benefits of vertical integration exist. First, vertical integration may be important for overall corporate strategy such as growth in revenue or diversification through expansion into equipment production. While this issue may be important, again analysis is difficult within the scope of this study, which attempts to examine behavior and policy in a particular industry. Second, vertical integration may improve efficiency, particularly in innovation, by allowing closer communication between stages of production. However, given the record of innovation and emphasis on understanding customer needs by nonintegrated-semiconductor firms, the advantage of vertical integration on this issue does not appear to be great. Third, vertical integration may provide a secure source of integrated-circuit supply for the equipment maker or a secure outlet for the integrated-circuit manufacturer. Having a secure source of supply may be a particularly strong motive for captive production. Having a partisan outlet to push integrated-circuits may also explain their makers' integration into consumer products. Fourth, vertical integration offers some protection from competitors for proprietary circuit designs. Finally, vertical integration may provide a source of cash to finance semiconductor expansion. This issue is discussed further in chapters 5 and 6.

Relationships among Strategies

Full-line firms must necessarily do some second sourcing, whereas specialized firms can concentrate exclusively on proprietary products if they choose. This aspect provides the most important overall link between a firm's product strategies and its pricing strategies. Proprietary-product suppliers are usually less aggressive on price than second sources. If production costs did not fall (due to the experience curve) with increased volume, the proprietary firm would have to achieve a higher price on average through the product cycle than second sources to cover the higher up-front development costs. Since production costs do fall, proprietary firms have some leeway in pricing. However, second sources must generally be aggressive in order to gain a portion of total product sales. A second source can be less aggressive only if few other firms are in competition. AMD, for instance, is said to have chosen the products it second sourced very carefully to avoid aggressive pricing and be as profitable as possible. Since second sourcing appears to make up a greater portion of full-line firms' product line than that of specialized firms, and since there is usually considerable competition, aggressive pricing will be a greater part of the full-line firms' overall strategy (interviews with industry executives).

As discussed earlier, the distinction between full-line and specialized firms is also an important aspect of marketing strategy. In addition, the case histories presented in the next section show the importance of delivery. New firms such as GMe and established firms attempting a major innovation, such as Sylvania, had trouble with delivery and were passed by other firms. National, in contrast, built more-complex devices to convince customers it could deliver less-complex devices. The importance of delivery in Intel's marketing strategy is underscored with its slogan, "Intel Delivers."

Since much of the performance analysis in chapter 5 must necessarily use data from established firms, the impact of delivery is difficult to analyze because the established firms have essentially mastered the production problems (for established products) that cause delivery problems. However, the distinction between full-line and specialized firms may also reflect potential delivery differences of importance in marketing. A full-line firm, for instance, can make more credible promises to deliver a new product than a specialized firm if the product is outside the latter's area of specialty.

Since many of the relationships among strategies that do exist relate to the distinction between full-line and specialized firms, this dichotomy forms a basis for classifying firms into strategic groups. We return to the concept of strategic groups in the concluding section of this chapter after reviewing strategies and competition in five important product areas in the following section.

Case Histories of Strategies and Competition

This section examines the history of the major products introduced by firms producing digital integrated circuits. Based on this historical review, we will then identify firm strategies and strategic groups (as were conceptually defined and explained in the framework volume). Understanding firm strategies, in turn, permits an analysis of firm and industry performance to be presented in chapter 5.

Considered from an aggregate, or total semiconductor-industry perspective, the fifteen-year meteoric growth in semiconductor-industry sales generally and digital integrated-circuit-market sales specifically might lead the casual observer to conclude that firms in this industry have found El

Dorado. However, closer examination reveals that many innovative digital integrated-circuit directions were tried and rejected in the marketplace. The firms committing most or all of their engineering resources to one technology and product approach have frequently been unsuccessful in their attempt to capture an important share of the growing digital integrated-circuit demand. The larger firms, pushing innovations on a broad range of technologies and product areas, have been successful overall in that they have experienced sales growth rates typical of the semiconductor industry. However, within these firms, some innovative directions succeeded while others did not succeed despite sometimes strongly held expectations to the contrary.

Competition in a high-technology industry often has the character of a race, with sizable rewards going to firms that gain a technological lead. This fact has been particularly true in the semiconductor industry. As Robert Noyce (1977, p. 68) has commented:

> A year's advantage in introducing a new product or new process can give a company a 25-percent cost advantage over competing companies; conversely, a year's lag puts a company at a significant disadvantage with respect to its competitors.[1]

In the semiconductor industry, such "technology races" have generally taken place within specific product fields. For example, semiconductor memories and microprocessors are two product fields in which races are now going on. Power transistors represent a more mature field in which the race is essentially over.

The technology-race concept provides a logical thread for sorting out the competitive process in semiconductors. The semiconductor industry produces a dazzling array of products—literally thousands of different types of parts. Some of these parts are new products at the threshold of technology. Others are mature products, where standardization has occurred and continued improvements are few. The nature of competition varies tremendously across these products, and for any given product, the nature of competition varies over time. Moreover, some semiconductor firms offer a broad product line whereas others are specialized. Any meaningful appraisal of competition must therefore focus on a limited number of products and firms. One way to accomplish this focus is to follow a particular technology race over time. We proceed by reviewing the innovative history and strategies of the major participants in five major product areas (early bipolar logic, custom LSI, LSI memories, consumer circuits, and microprocessors). We will look at some of the product ideas and firms that succeeded and failed in each of these areas to develop a picture of the strategies typically followed by firms.

The Bipolar-Logic War

The first big demand for digital integrated circuits developed in the middle to late 1960s. A variety of bipolar families, or kits of electrically compatible building-block integrated-circuit products, were developed. Through selecting and interconnecting products in the family, the integrated-circuit customers design and build their own equipment (for example, a computer terminal). These families competed for the digital demand, with the TI-designed 54/74 TTL emerging as the eventual winner in terms of widest usage, market acceptance, and growth.

In the early 1960s, however, this outcome was neither planned nor widely imagined (save perhaps by some individuals at TI). The first proprietary integrated circuits introduced were Fairchild's Resistor-Capacitor-Transistor Logic (RCTL) family and TI's Direct-Coupled Transistor Logic (DCTL) (*Electronics*, 14 November 1966). Sales to the military developed early because integrated-circuits allowed for more compact designs in military hardware. TI's integrated circuits were designed into the Minuteman Missile, an early triumph for integrated circuits over discrete-device logic designs (Sickman 1966).

These early product families had design weaknesses and were quickly supplanted first by RTL and then by DTL. Fairchild pioneered RTL but after some initial success it was supplanted by DTL because of problems with noise immunity, that is, the resistance of the integrated-circuits' input section to spurious or false signals stemming from electrical "noise" in the system. Fairchild, TI, and Signetics had proprietary DTL approaches (*Electronics*, 6 March 1967). Fairchild's 930-series DTL emerged as the industry leader in the period from 1965 to 1967. Fairchild achieved its leadership by seeking to penetrate the computer and industrial markets through aggressive price cutting (Golding 1971, pp. 161-162). Perceiving the emerging dominance of the Fairchild 930 DTL design ten suppliers were second sourcing the 930 series by August 1967 (Golding 1971, p. 162). To penetrate a market with a new design, an integrated-circuit-family innovator must have at least one second source or alternative supplier. This requirement for second sources was a convention adopted by purchasers of semiconductors after having been frequently "burned" by sole sources delivering products late or not at all in the early 1960s. (Second sourcing as a strategy is discussed later in this chapter.)

Fairchild's 930-series DTL did not maintain its dominance for long, however. While DTL was at its peak, Sylvania was pushing its proprietary Sylvania Universal High-Level Logic (SUHL), which was the first of the TTL families. (Interestingly Fairchild invented TTL in 1964 but did not market a TTL family until 1967.) The biggest advantage of TTL was its fast switching speed (*Electronics*, 18 September 1967, pp. 179-182). TI quickly followed suit with its own proprietary TTL line, the 54/74 series, and the Sylvania versus TI battle was on. Although it seems doubtful that anyone could have foreseen it, TTL demand grew explosively (*Electronics*, 18 September 1967). Second sources quickly lined up behind Sylvania (Motorola, Raytheon, Philco-Ford, and Westinghouse) and TI (National and Sprague). Signetics, as it did with DTL, went with its own proprietary TTL design (Designers-Choice Logic) (*Electronics*, 8 January 1968 and 18 September 1967). Fairchild, however, languished behind Sylvania and TI and did not introduce its proprietary 9000-series TTL until 1967 (*Electronics*, 18 September 1967). Sylvania's design approach differed from TI's in that Sylvania focused on improved design of the components of its integrated-circuits while the TI focus was to cram more and more logic on the integrated-circuits (*Electronics*, 18 September 1967). This "cramming" of more functions on a chip led to newer TTL products that could perform complete computer functions (for example, adders, decoders, encoders, 4-bit serial memories) and were called medium-scale-integration (MSI) devices.

The rapid rise in demand for TTL and low yields by Sylvania (*Electronics*, 18 October 1969) enabled TI to become the leading producer of bipolar logic. However, a principal cause of the stampede to 54/74 was the role played by National. National was taken over in 1967 by a team from Fairchild led by Charles Sporck and went looking for a line of digital integrated-circuits to second source and chose 54/74. To penetrate the market, they first built MSI devices to convince buyers that they could deliver (*Electronics*, 13 October 1969). Delivering MSI convinced buyers that National could deliver the 54/74 TTL devices (less complex than MSI) (*Electronics*, 13 October 1969). Then National began an aggressive price-cutting campaign, which cemented the 54/74 victory. Sylvania lost out because it could not deliver, and the firm stopped making semiconductors in 1970.

In 1970 TI introduced a major innovation that greatly increased the speed of bipolar logic. Termed "Schottky TTL," this innovation was incorporated in an improved-performance line of 54/74-type circuits. Three years later TI developed "low-power Schottky," which offered a speed advantage as well as low power consumption.

We now can make some conclusions about firm strategy. First, succss in the semiconductor industry requires four characteristics: a product well designed for customer acceptance, ready second sources, aggressive pricing, and ability to deliver to the developing market. The absence of any one of these factors will doom the innovator, and perhaps the innovation, to failure. Sylvania could not deliver TTL and left the business. TI succeeded in part because of National's aggressive second sourcing. Fairchild succeeded in DTL with aggressive pricing. DCTL, RCTL, and RTL sales never grew because of poor customer acceptance and obsolescence created by improved designs.

The second conclusion we can draw is that innovators in one phase of the battle can be followers in the next phase. Fairchild's history in this period best illustrates the point. Successful innovation of DTL was quickly followed by a complete lapse in the TTL area, a design concept that Fairchild had, in fact, invented. Fairchild had a number of problems at the time that contributed to the lapse. These included poor yields, too many new products in design, late delivery, and internal control problems. In addition, Fairchild's strategy was to produce only proprietary designs and they refused to second source competing lines (Golding 1971, p. 163). Finally, Fairchild apparently did not expect TTL to replace DTL as widely as it did; the firm questioned how widespread the demand for high-speed circuits in small computers and industrial equipment would be (interview with former industry employee). Successful semiconductor-industry strategy requires a willingness to change direction quickly when early intuitions prove wrong. Few firms exist that one can unambiguously classify as innovators and still fewer of these are successful (for example, note Sylvania's demise).

The third conclusion is that room is available for aggressive second sources in the role of making an innovation a success. National's role in TTL illustrates the point that without its aggressiveness, Sylvania or the emerging Fairchild 9000 series might have eventually triumphed.

Finally, the fourth conclusion is that being first with an innovation does not guarantee success. Sylvania's lack of success with SUHL TTL because of yield and delivery problems provides a good example of this point.

Custom LSI: The Boom that Never Happened

While the bipolar-logic war was in full heat, semiconductor firms were also making plans and spending money to establish their position in what they believed to be the next generation of digital integrated circuits: custom LSI.

The definition of LSI varies from time to time and from firm to firm because the term LSI is more a marketing term than a precise technical one. In the 1966 to 1968 period an LSI circuit was usually considered to be one of about 100-gate (about 300 transistors) density and containing two layers of metal (for instance, aluminum) separated by silicon oxide (for instance, glass) interconnecting the transistors on the chip. Mature products marketed in that period (and for some time afterward, as things turned out) utilized only single-layer metal interconnections.

The bipolar-logic area previously described was dominated by standard off-the-shelf parts. The various part numbers within a family (for example, the DTL family) were purchased by the equipment manufacturer who wired them together to achieve equipment design objectives.

A custom circuit is manufactured by a semiconductor firm that works closely with the customar to design and build a circuit optimized for the customer's specific design. Standard parts, in contrast, are designed for wide usage in a variety of equipment designs. Although not universally held, the general semiconductor-industry consensus in the late 1960s was that the future of digital integrated circuits lay in custom LSI. The need for standard parts was expected to be small (*Electronics*, 20 February 1967). The principal reason LSI was expected to be dominated by custom rather than standard products was that most semiconductor experts believed that the innovating equipment maker would not want to find that the complex integrated circuits, which determined the performance of its product, were readily available to its competitors (*Business Week*, 27 March 1971). (Although not expressed until 1971, this sentiment was held four years earlier.) With early bipolar logic the equipment maker could differentiate itself from competitors by cleverly interconnecting standard integrated circuits. In a design where LSI was to be used, most of the logic needed was now packed into a few integrated circuits and therefore much fewer product-performance-differentiating options were open to the equipment manufacturer, unless the LSI used were circuits custom designed to optimize system performance.

In 1967 the question was not whether custom LSI was the future demand, but what was the best route to the custom LSI that customers would want? Firm strategies can be differentiated according to technical approach. Fairchild and TI were not yet focused on bipolar or MOS LSI approaches but were engaged in a vigorous public-relations contest about the most economical way to build custom LSI circuits (both bipolar and MOS).

Both TI and Fairchild believed that the best approach to LSI was to standardize all aspects of the circuit design except the final-mask (or metal-mask) stage. In other words, both Fairchild and TI intended to build wafers for inventory containing logic gates. The circuit would be customized by pulling the wafers from inventory and processing the wafers through the final wafer-fabrication-production stage (see chapter 2), which would connect the logic gates on the wafer into the customer's unique circuit configuration. Computer-aided design, assisting the translation of the customer's logic design into an optimal logic-gate-interconnection plan, was considered essential (*Electronics*, 20 February 1967). The R&D hardware and software investment for such computer-aided design systems was substantial, but both Fairchild and TI were making the effort. The TI approach to custom LSI was called the Discretionary Wiring technique, and the Fairchild approach was trademarked Micromatrix.

The other innovative custom LSI direction that firms were following in the middle and late 1960s focused mainly on MOS. The usual custom MOS approach did not rely on computer-aided design of two-layer-metal circuits. The innovators' strategy was to accept orders for custom circuits designed by standard single-layer-metal techniques. The circuits were labeled LSI because MOS technology allowed the designer to pack more function density into a given chip area.

The pioneer of this approach was GMe, a 1964 Fairchild spinoff (*Electronics*, 14 October 1965), which was taken over by Philco-Ford in 1966. Other small firms such as AMI, General Instrument, and North American Rockwell joined in the custom MOS-LSI race in the late 1960s. Of the large firms, Fairchild and National were active in MOS, but National was quite cool toward the custom LSI market (*Electronics*, 13 October 1969). Fairchild expected to integrate its MOS and Micromatrix efforts, but devices actually shipped in the 1960s were designed using standard approaches. TI did not begin custom MOS sales until 1970 (*Business Week*, 25 April 1970). Signetics and Motorola were also not active in the custom MOS-LSI effort of the 1960s. Signetics believed that the important LSI products would be bipolar (*Electronics*, 9 January 1967). Motorola was not an important innovator nor was it as aggressive a second source as was true of its strategy in other digital integrated-circuit-product areas. However, Motorola acknowledged that MOS was likely to prove important (*Electronics*, 8 August 1966).

Early forecasts for MOS generally and by inference, custom MOS, were bullish (*Electronics*, 14 October 1965). However, total MOS sales were not very large until 1970. AMI emerged as the MOS leader in 1969 (*Business Week*, 25 April 1970).

As late as 1971 the industry still expected custom as opposed to standard LSI to prevail (*Business Week*, 27 March 1971). However, some notable defectors began to emerge. National and a small firm, Electronic Arrays, believed that standard circuits, not custom, was the way to proceed. Intel, a new entrant in 1968, pushed standard catalog LSI memories, both MOS and bipolar. Other firms such as General Instrument were hedging bets in 1970 by offering both customs and standards but the standards offered were modifications of custom designs. AMI and North American Rockwell still pursued custom business (*Business Week*, 25 April 1970).

By the early 1970s such concepts as Micromatrix and Discretionary Wiring had faded into oblivion (*Electronics*, 10 January 1974). LSI was largely MOS LSI in the 1970 to 1974 period in terms of largest absolute value and growth rate of sales dollars.

LSI end-user directions became clearer also in the 1970-1974 period. Two principal MOS-LSI product areas had emerged as dominant: calculators and memories. Watches were also an important emerging area of demand. Intel was the most important memory innovator of the period and pushed the memories toward standard circuits. Intel's 1103 1K RAM was a standard product and the first large-volume memory circuit. The integrated-circuit manufacturer of calculator chips also faced the custom LSI versus standard LSI choice. By early 1974 most firms were talking of the latter choice, as the market had distinctly swung to standard products (*Electronics*, 10 January 1974). Bipolar LSI, fairly dormant in the big-dollar early 1970s LSI race, had begun to make a comeback. However, the bipolar comeback was for standard products also.

In the fast-paced world of integrated-circuit product and market innovation, concepts and generalizations about firm strategy are difficult to make, but we believe that the following are accurate.

The 1964 to 1974 era might be called the era of custom LSI in the sense that many firms believed that custom LSI products would constitute the largest portion of the next generation of digital integrated-circuit sales dollars. In the first half of this decade the giants of the industry (TI, Fairchild, and Motorola) were preoccupied with the bipolar-logic war described earlier and lost some of the innovative momentum to small firms. In turn, the smaller firms, unable to capitalize on their lead, lost it to the large firms in the second half of the custom LSI decade. Intel, the notable exception, was never really in the custom LSI business. AMI is an example of an exception that enjoyed some moderate success from custom LSI in the period. However, by the second half of the decade the market envisioned by both the large- and the small-firm innovators clearly was never going to meet expectations. The TI (Discretionary Wiring) and Fairchild (Micromatrix) approaches were unsuccessful. The only real (large-dollar) success from 1964 to 1974 enjoyed by custom LSI occurred in sales to calculator original equipment manufacturers (OEMs). However, calculator integrated-circuits moved later to standards sold to OEMs or were manufactured internally by vertically integrated equipment makers.

We mentioned before that four ingredients were necessary for successful innovation in the digital integrated-circuit industry (design for customer acceptance, ready second sources, aggressive pricing, and ability to deliver). In retrospect it seems clear that custom LSI really offered the buyer only one of the four: design for customer acceptance. Custom LSI offered none of the advantages of wide second sourcing, ready availability, and falling prices characteristic of standards. The latter three advantages of standards resulted not so much from deliberate firm strategy but rather stemmed from the fact that the forces of competition operate more vigorously in standard circuits.

The custom LSI history reinforces a conclusion drawn from the bipolar history: even in a booming industry, many if not most innovations can prove unsuccessful. Innovative strategies followed by even the most technologically sophisticated firms often turn out to be wrong. Perhaps the two most successful LSI innovators of the period, Intel and National, might be called successful simply because they resisted the siren call of custom LSI and saw correctly that the big-dollar future for LSI lay in the market for the standard calculators, memories, microprocessors, and consumer circuits discussed in the next sections.

The Memory Race

Semiconductor memories are used to store information. Their primary application is in computers and related equipment. Although several general types of semiconductor memories are available this discussion is limited to random-access memories, or RAMs. A random-access-memory design means that any piece of information can be retrieved directly without first searching through all the information stored ahead of it or behind it. In contrast, with a serial or shift-register memory all the information may have to be retrieved before the desired piece is found. A magnetic stereo tape is an example of a serial memory. You may have to play or "fast forward" by several pieces of music you do not want to hear before reaching (that is, retrieving from the memory) the desired piece.

RAMs constitute the most important type of semiconductor memory in terms of sales. RAM sales have grown rapidly, from $36 million in 1972 (*Electronics*, 10 January 1974) to more than $500 million in 1979 (*Electronics*, 4 January 1979). At the same time, prices have fallen and storage capacity per integrated circuit has been improved. Real performance in terms of information-storage capacity has therefore risen much faster than the sales figures indicate.

Semiconductor memories can be fabricated using either MOS or bipolar technology. MOS technology offers relatively low cost, whereas bipolar technology has an advantage in speed of information retrieval. MOS has historically been the sales leader in RAMs by a considerable margin. Prior to the 1970s magnetic cores were used to provide random-access memory in electronic computers. Semiconductor memories have now replaced magnetic cores in virtually all applications, and rapidly falling prices have led to vastly increased use.

The performance of semiconductor memories has improved over time in four dimensions. First, storage capacity per integrated circuit has quadrupled three times from 1,000 storage cells (or bits) in 1970 to 64,000 today. Second, the price per bit of storage has fallen from about one cent per bit in 1970 to five-hundredths of a cent in 1979 due to the experience curve and increasing storage density per integrated circuit. The experience curve and increasing circuit density's relation to price were explained in chapter 2. Third, the time it takes the memory to retrieve stored information (access time) has decreased. Fourth, the ease of use of integrated-circuit memories has improved in two principal respects. They are now easier to use in conjunction with standard TTL logic, and some memory products sold do not require that their stored information be periodically refreshed to correct for stored-information loss due to internal signal fading.

Semiconductor integrated-circuit memories have been produced since the mid-1960's, but the race to produce a low-cost, high-performance memory did not begin in earnest until 1968. By the late 1960s it was clear to many people in the semiconductor industry that a huge market would become available if semiconductor memories could be developed to replace the magnetic cores then used in electronic computers. The cost of fabricating conventional bipolar integrated circuits had fallen sharply, and MOS technology was showing considerable potential as a magnetic-core replacement approach.

Unclear at this time was which technology, bipolar or MOS, had greater potential for semiconductor memories. Bipolar memories were on the market, but they were more expensive than magnetic cores for most applications. MOS technology was unproven, but it held the prospect of much lower production cost than bipolar.

Given this technological and market opportunity, several new companies were formed in 1968 to develop semiconductor-memory products. Two of these new firms, Intel and Mostek, ultimately became the sales and technology leaders. Other new entrants included Advanced Memory Systems, Computer Microtechnology, Semiconductor Electronic Memories Incorporated (SEMI), and Cogar Corporation (Bylinsky 1973).

The established semiconductor firms were also developing semiconductor-memory products. For example, in 1968 Motorola announced its first MOS product, a 64-bit RAM (*Electronics*, 30 September 1968, pp. 189-190). Nevertheless, apparently the established firms failed to make major commitments to semiconductor-memory technology at this point. This failure created opportunities for the new entrant firms.

By early 1970, at least eighteen companies were making or planning some form of semiconductor memory. The first major breakthrough was in 1970 with the introduction of the first 1K (1,000 bit) RAM. Intel pioneered the development of polysilicon MOS technology, which was instrumental in breaking the technological barrier to production of the 1K RAM. Advanced Memory Systems, however, was first to market a 1K RAM; Intel followed three months later. Intel's device was not as fast as the Advanced Memory Systems' unit but was easier to use (Altman 1972). Also important, Intel waged an effective advertising and promotion campaign that stressed its ability to deliver. Advanced Memory Systems, however, had financial trouble and was unable to mount an effective marketing strategy (Sideris 1973). Intel, therefore, quickly took the lead in 1K-RAM sales.

The success of Intel's 1103 led to a response by leading firms in the industry. Fairchild, TI, and other major semiconductor producers approached Intel to request licensing as second sources. Intel refused to license these major U.S. firms. Instead, Intel licensed a small Canadian firm, Microsystems International Ltd., as a second source (Bylinsky 1973, p. 184). Second sourcing, discussed earlier in this chapter, is common industry practice. A proprietary product innovator needs a second source to win buyer confidence in reliable delivery. The innovator can either select and license an obscure firm or a major competitor as a second source. The choice of a major competitor increases buyer confidence, while the choice of a small "fringe" firm minimizes the toughness of price competition in large-quantity sales negotiations. In addition to licensed second sources, other firms can second source by designing their own unit that will meet the performance specifications of a competitor's fast-selling new proprietary product.

Advanced Memory Systems responded to Intel's 1103 by introducing a faster 1K RAM and signing up TI and Motorola as second sources (Altman 1972, p. 68). TI, however, failed to get into production either this second-source product or its own proprietary memory. Instead, TI announced that it would "leapfrog" technology and focus its efforts on a 4K memory.

As Intel's 1103 became firmly established as the dominant product, American Microsystems and Fairchild both came out with their "own versions" of the 1103. Intel had already won the first round of competition.

In 1972, Fairchild announced the first 1K bipolar semiconductor memory (Altman 1972). Although more expensive than MOS memories, Fairchild's device was much faster. This speed advantage gave Fairchild a major position in high-speed RAMs. For the next several years Fairchild continued to build on its strength in bipolar technology, rather than compete directly with the newly established firms in MOS RAMS.

In 1973 the second round began when 4K MOS memories were introduced by Intel, TI, and Mostek. The three firms each vied for position to have their design become the dominant supplier. Intel announced its 4K memory in July and made its first shipment in September (Bylinsky 1973; Riley 1973). Microsystems International Ltd. of Canada was again licensed by Intel as a second source. TI and Mostek both announced their 4K memories in August (Bylinsky 1973).

Not all firms were eager to be first to market, however. Advanced Memory Systems and National announced that they would hold back on development and production of a 4K memory until an industry standard clearly emerged (Sideris 1973; Wolff 1973).

In October 1973, Intel decided to modify its new 4K memory to make it compatible with TI's product. Shortly thereafter, TI encountered production problems with its 4K memory and was unable to meet supply commitments (*The New York Times,* 13 May 1979, p. F-1). At the same time,

Mostek's 16-pin design was gaining favor over the 22-pin designs of Intel and TI. By 1974 Mostek's 4K memory had emerged as the standard for most of the industry (*The New York Times*, 13 May 1979, p. F-1). With Mostek's memory as a standard, new 4K memories were introduced by National, Motorola, American Microsystems, and others (Wolff 1973; *Electronics*, 13 June 1974).

In 1975, RCA announced its first 1K RAM (*Electronics*, 16 October 1975). RCA's product was unique in that it utilized "CMOS-Silicon-on-Sapphire" technology, which had been pioneered by RCA in the early 1970s. CMOS-Silicon-on-Sapphire technology offers the advantage of low power consumption and relatively high speed but at a high price compared with conventional MOS. RCA's memory met the needs of a small but specialized group of military customers who valued these technical attributes strongly. RCA's strategy was thus to exploit its proprietary technology to serve a specialized market niche rather than compete directly in the mainstream of semiconductor memory sales.

In 1975, Fairchild announced its first 4K bipolar memory (*Electronics*, 16 October 1975). This bipolar memory was two to three times faster than comparable MOS memories, and it sold at a correspondingly higher price.

Several months after Fairchild's announcement, TI began production of its first 4K bipolar memory (*Electronics*, 28 October 1976). As the largest and most diversified firm in the industry, TI chose to produce both MOS and bipolar memories.

During 1976 and 1977 semiconductor-memory sales grew rapidly. Mostek, Intel, and TI had become firmly established as the major producers of 4K dynamic MOS memories, with a combined market share of more than 80 percent according to *Electronics* (Altman 1977, p. 88). However, many other firms had entered as alternate sources or with their own proprietary 4K-memory designs. By early 1977, Motorola was second sourcing Mostek, and Fairchild had plans to do so. Motorola also offered its own proprietary design. AMD, Signetics, Electronic Arrays, and RCA were alternate sources to TI. National Semiconductor had developed two proprietary designs, working with American Microsystems as a second source. Japanese firms, Nippon Electric Company, Fujitsu, and Hitachi, were also selling 4K RAMS in the United States, although in small quantities (Altman 1977, p. 88).

Intel, Mostek, and TI each announced 16K dynamic memories in 1975, but by early 1977 relatively few units had been shipped (Altman 1977, p. 88). Mostek and Intel developed "double-polysilicon" technology for their 16K RAMs; TI did not, and later had to redesign its part using this technology (Altman 1976, p. 74). Mostek's design became the industry standard. All three firms encountered production problems, however, which enabled Japanese firms to gain a sizable share of 16K sales. By mid-1979 sixteen companies were producing 16K dynamic RAMs, and the Japanese producers accounted for 42 percent of the business (*Business Week*, 3 December 1979, p. 86).

The largest semiconductor memories introduced to date are 64K models. Fujitsu was the first firm to introduce this product in the United States. In late 1978, TI followed with a similar device. By mid-1979, two other firms, Motorola and Hitachi, had units in production, and Mostek, Intel, and National indicated that they also planned to produce this part (*Status*, 1979, p. 2-12).

Intel, the early leader in MOS-memory technology, was not among the initial producers of the 64K memory. By 1979, Intel had chosen to reduce its efforts considerably in the dynamic MOS-memory field, concentrating instead on other, more profitable products. Product standardization and entry into the dynamic-memory field had led to increased price competition and decreased profit margins. Moreover, the direction of new technological developments had become increasingly predictable. Intel and other U.S. firms found it profitable to reallocate their dynamic MOS-memory production lines to other, higher-margin products. To cover existing supply commitments, they purchased dynamic memories from Japanese suppliers (*Business Week*, 3 December 1979, pp. 85-86).

The technology race in semiconductor memories illustrates the evolution of one branch of digital integrated-circuit technology over time. The advent of semiconductor-memory technology, as with earlier digital integrated-circuit technologies, created opportunities for entry by new firms. During the course of the technology race, production costs fell and a series of standard products emerged. As standardization increased, technological advances became easier to predict, the total number of producers grew, and foreign producers expanded their market share. These changes are all consistent with the product life-cycle model. [For an introduction to the product life-cycle-model concept, see Wells (1972).]

The technology race in semiconductor memories also illustrates the variety of strategies pursued by semiconductor firms. Intel, newly formed in the late 1960s, quickly gained leadership in technology and sales. Intel maintained that leadership until segments of the market began to mature in the late 1970s. TI, the largest of the established firms, tried to move quickly into the new MOS field. TI initially failed in this attempt but ultimately became a major producer. Fairchild, another large, established firm, chose to build on its expertise in bipolar rather than compete directly with the new MOS firms. Motorola was less of a technology leader, hedging instead between proprietary and second-source products. Thus in memories, as in bipolar logic, diversity of strategies was important. Those product directions that succeeded met four criteria: customer acceptance, second sourcing, aggressive pricing, and credible delivery commitments.

Microprocessors

A microprocessor is a component that contains all the elements of a computer's central-processing unit (*Fortune*, November 1975, p. 90). Though some microprocessors consist of two or more integrated-circuits, most are single integrated-circuit devices. The microprocessor is usually combined with memory and input-output integrated-circuit devices to form a "microcomputer." Some microcomputers also contain the microprocessor, memory, and input-output circuits all in a single integrated circuit. Four general product segments of microprocessors have been developed, called 4-bit, 8-bit, 12- and 16-bit, and bit slice. The number of bits determines the microprocessor's word length, which in turn determines, in part, the precision and complexity of calculations possible within a given microprocessor. Except for bit-slice devices, virtually all microprocessors are manufactured with MOS technology. The necessary spacing to allow for heat dissipation currently makes difficult the manufacture of an entire microprocessor on one chip with bipolar technology. Therefore, if the high speed of bipolar technology is desired, several bit-slice units are usually coupled in parallel to form a microprocessor (Vacroux 1975, p. 36).

Intel invented the first 4-bit microprocessor in 1971 (*Fortune*, November 1975; *Financial World*, 15 March 1977, p. 16) and began deliveries in the second quarter of that year (*IC Master*, 1977; *Datamation*, December 1974). Other firms were slow in entering the microprocessor area. Only Rockwell International and National introduced their own 4-bit designs within a year (*IC Master*, 1977; *Fortune*, November 1975, p. 137). Fairchild also introduced a 2-chip microprocessor, primarily for use in calculators (*Electronics*, 1 March 1973).

As the microprocessor gained customer acceptance, other firms began development of their own designs. Meanwhile, Intel charged ahead with more advanced products. In the first quarter of 1972, before other firms had introduced their own 4-bit products, Intel began deliveries of an 8-bit microprocessor, and in 1973, still before most firms had yet entered the 4-bit area, Intel introduced the 8080, an 8-bit microprocessor that was 20 times faster than Intel's 4-bit device (*IC Master*, 1977; *Fortune*, November 1975). One source estimates that Intel's 4-bit and 8-bit products had 99 percent of microprocessor sales in March 1973 (*Electronics*, 1 March 1973).

Though Intel continued to hold an 80 percent share of the microprocessor product area in 1974 (*Financial World*, 15 March 1977), it faced increasing competition. Only Motorola challenged Intel in 8-bit microprocessors in 1974 with deliveries of its 6800 in the second quarter (*IC Master*, 1977). However, in the second quarter General Instrument entered with a 16-bit microprocessor, and Toshiba entered with a 12-bit microprocessor.

Monolithic Memories introduced the first bipolar microprocessor (a TTL 4-bit-slice design) in 1974 (*Electronics*, 3 August 1978), though Intel quickly followed with a TTL 2-bit-slice product in the third quarter (*IC Master*, 1977). AMD followed with its own TTL 4-bit-slice microprocessor in the third quarter of 1975 (*IC Master*, 1977). While Monolithic Memories provided only a bit-slice processor product, Intel and AMD provided whole families of support circuits for their microprocessors and design aids to make their use easier. These additional products brought bit-slice microprocessors to customers' attention (*Electronics*, 3 August 1978). AMD seems to have been particularly successful with its product; Fairchild, Monolithic Memories, Motorola, National, Raytheon, and Signetics eventually second sourced the AMD device (*Electronics*, 3 August 1978).

In 1975 TI finally entered the microprocessor area with a 4-bit P-MOS microcomputer, and Fairchild introduced an 8-bit N-MOS device. MOS Technology, RCA, Signetics, Scientific Microsystems, and Western Digital also introduced proprietary designs (*IC Master*, 1977). Many 8-bit second-source products were also introduced. The 8080 was becoming the industry standard, and many firms second sourced the product: TI, Nippon, AMD, Mitsubishi, Oki Electric, and Microsystems International (*IC Master*, 1977; *Fortune*, November 1975). Only Microsystems International had a second-source agreement with Intel. Motorola's 6800 was probably in second place, and AMI began producing the 6800 with a second-source agreement with Motorola (*IC Master*, 1977). Synertek began second sourcing MOS Technology's 6500 and Mostek Fairchild's F-8 (*IC Master*, 1977; *Electronic News*, 13 October 1975).

The mass entry into microprocessors in 1975 forced firms to cut their prices. Motorola was the first to cut prices in the 8-bit area; in September 1975 it reduced distributor-level prices as much as 60 percent on the 6800. Intel followed in October with price cuts of up to 50 percent on small-quantity orders of the 8080 and peripheral devices. Other suppliers also began cutting prices (*Electronic News*, 8 March 1976, p. 56).

In 1976 Intel continued as the microprocessor sales leader. Motorola with its 6800 increased its share of 8-bit-microprocessor sales from 10 to 15 percent in 1976 to 20 to 25 percent in 1976 (*Electronics*, 11 November 1976). TI, Zilog, National, Rockwell, RCA, and Motorola introduced new proprietary products in 1976, and National, Motorola, Signetics, Intersil, Mostek, Electronic Memories and Magnetics (EMM), Harris, and Raytheon began deliveries of new second-source products (*IC Master*, 1977). Noteworthy among the new products were Zilog's improvement of the Intel 8080 and TI's NMOS 16-bit microprocessor.

Intel continued to hold on to its microprocessor sales leadership in 1977 with 60 to 65 percent of total microprocessor sales and according to *Finan-*

cial World (15 March 1977) it had the most profitable microprocessor operations as well. Motorola continued in second place with 15 to 20 percent of microprocessor sales, and TI was third with 10 to 15 percent of total sales. Fairchild was also successful with its 2-chip F-8 microprocessor, the leading microprocessor in terms of central processing unit (CPU) sales in 1977 (Osborne and Associates 1978, chapter 2). However, Mostek introduced a major improvement on the F-8 and Fairchild became a second source to Mostek. In 1977 Fairchild also began deliveries as a second source of Motorola's 6800 (*IC Master*, 1977; Osborne and Associates 1978).

Because of the fierce competition in the 8-bit area, particularly from new, high-performance microprocessors such as Zilog's Z80, Intel also introduced higher-performance 8-bit products. Competition and the experience curve continued to drive down microprocessor prices. In the third quarter of 1975 the price of the Intel 8080A was $110, in the first quarter of 1976 it was $40, and in the first quarter of 1977 it was down to $20 (*Financial World*, 15 March 1977).

In 1978 and 1979 the microprocessor manufacturers aimed their technological advances at the 16-bit microprocessor. TI, General Instrument, National, and Western Digital were selling older 16-bit designs in mid-1978 when Intel introduced the 8086, the first of a new generation of 16-bit microprocessors (*The Economist*, 6 October 1979). Zilog introduced its 16-bit microprocessor in early 1979 (*Electronics*, 15 March 1979), and Motorola and National announced their versions in late 1979 (*The Economist*, 6 October 1979).

Though Intel now has the largest share in the new 16-bit segment, it is not as far ahead as it was in the 4-bit and 8-bit segments. Zilog's 16-bit product is generally considered to have superior performance. While Intel was concerned about maintaining electrical compatibility with its 8-bit units, Zilog was relatively more concerned with achieving maximum performance out of a 16-bit microprocessor (*Electronics*, 23 November 1978; *The Economist*, 6 October 1979). However, Intel hopes to introduce an improved version of its 16-bit product in early 1980. Zilog is also working on an improved version of its 16-bit product, though it will not introduce it before late 1980 (*The Economist*, 6 October 1979).

Additional evidence of the keen competition in the 16-bit segment has been the jockeying for second-source agreements. When Intel introduced the 8080, it felt little need for the support of a second-source agreement. Though many firms eventually second sourced the 8080, Microsystems International, a small Canadian firm that posed little threat to Intel, was the only firm to obtain an agreement with Intel early in the product's life. However, in the 16-bit area Intel was eager to find a second source. AMD, a second source for Intel's 8-bit products, made an agreement with Zilog in September 1978 to second source its 16-bit product. A month later Mostek, which competed in the 8-bit segment with its own device and a second-source version of the Zilog product, announced its agreement to second source Intel's 16-bit product (*Electronics*, 23 November 1978).

Though the recent advances have been primarily in the 16-bit area, sales in the other areas have continued to grow. In fact, 4-bit-microprocessor volume has increased faster than total microprocessor volume (*Status*, 1979, p. 3-1), and 4-bit microcomputers have been particularly successful. Though inferior in price and performance, old designs such as the Intel 4004 continue to bring sales because customers cannot justify redesigning their products around the newer components. Nevertheless, the decreasing prices of 8-bit microprocessors are making the old 4-bit products obsolete (Osborne and Associates 1978, chapter 1).

While firms compete on the basis of performance in the 16-bit area, they compete on the basis of price in the 4-bit area. In addition, few firms remain in the 4-bit area. In 1978 TI and Rockwell each had about 45 percent of the low-end microprocessor sales (*Electronics*, 3 August 1978, p. 50).

Strategies followed by firms in the microprocessor race are similar to those found in the other standard product races. Several important points are worth emphasizing. First, timing is an important ingredient in successfully implementing a major innovation but not the sole ingredient. Rockwell and National, for instance, were in the microprocessor race early. If they had been more successful, the history would read differently. Motorola got into the race later, but came very close to beating Intel with the first really popular design. Motorola's 6800 was introduced only a short time after Intel's 8080. Thus at the critical juncture, Intel's early lead boiled down to a few months' advantage. Yet even those few months were crucial to the success of one firm versus the other.

Second, behavior in 4-bit microprocessors illustrates strategy with respect to older semiconductor products. As other firms withdraw to concentrate on newer segments, only a few firms are left in the older areas, giving the remaining firms a limited (by competition from other products) power over price and consequent potential for profits.

Finally, once again, a spinoff firm emerges as a contender in the race. Zilog, a firm that specializes in microprocessors, was formed in 1977 by key microprocessor people from Intel (*Financial Times*, 13 June 1977). However, in contrast to spinoffs in earlier periods, Zilog went to a very large company, Exxon, for financing.

Consumer Products

When the MOS-LSI technology finally matured in 1969 and 1970, the high function density per circuit and low cost-per-function MOS potential was tailored to such consumer market applications as calculators and watches.

MOS-LSI firms penetrated these markets, formerly dominated by stable firms selling mechanical designs, with aggressive pricing. Besides aggressive pricing, some integrated-circuit firms adopted a strategy of integrating forward and marketing products under their own label. Others continued in the traditional form of OEM sales. We will now review the specifics of the calculator and watch strategies followed.

Calculators. Japanese calculator manufacturers were the first to exploit the consumer electronic-products area in the late 1960s and 1970s. Because of their expertise in LSI circuits, a number of U.S. firms obtained long-term contracts with Japanese firms for the production of calculator circuits and the provision of technical assistance (*Japan Economic Journal*, 5 May 1970, p. 10). Rockwell provided technical assistance for Sharp, General Instrument provided chips for Sanyo, TI provided circuits for Canon (*Japan Economic Journal*, 5 May 1970, p. 10), and Fairchild supplied circuits for Casio (*Japan Economic Journal*, 3 November, 1970, p. 10). Besides providing a large demand for MOS circuits, these contracts provided the experience U.S. firms used later in the domestic calculator market. In particular TI, a latecomer to MOS technology, benefited from its contract to supply circuits for Canon's calculator, the first all-electronic, pocket-size, portable calculator (*Business Week*, 25 April 1970).

Bowmar, a U.S. firm, introduced the first pocket calculator in 1971 (*Business Week*, 24 August 1974). Many other firms such as Commodore quickly joined Bowmar as calculator assemblers, while the semiconductor firms remained as component suppliers. TI introduced its first three calculators in September 1972 with aggressive pricing. The price of its first calculator was $149. Bowmar responded with a price cut of $30 from its $179 list price. TI then cut prices another $30 (*Business Week*, 24 August 1974).

By late 1973 TI was the leading calculator producer. National and Rockwell also produced calculators; National had its own brand, Novus, while Rockwell produced for private labels. In 1974 TI, National, and Rockwell accounted for over 50 percent of calculator sales, though 45 U.S. companies produced calculators (*Business Week*, 24 August 1974).

TI continued its aggressive pricing strategy. During a 60-day period in 1974 it cut prices on low-end models by up to one-third. In 1974 TI also directed its marketing strategy at shortening distribution channels through direct-mail sales (a method used earlier by Hewlett-Packard) and even through its own retail outlets. Rockwell remained with private-label products but developed a broader line of products than TI to simplify retail selling (*Business Week*, 24 August 1974).

The effect on prices of such intense price competition and expansion in supply from new entrants hit many firms hard. Twenty-nine North American and Japanese companies left the calculator business in 1973 and 1974.

Both Bowmar and Commodore stuffered losses and attempted to remain price competitive by integrating backward into semiconductor production. Japanese firms were unable to keep up. Their share of U.S. calculator sales in 1974 dropped to 25 percent from 85 percent in 1971.

In 1974 National, TI, Rockwell, General Instrument, Electronic Arrays, Western Digital, Mostek, and MOS Technology were among the component suppliers (*Electronics*, 31 October 1974). Calculator-circuit prices dropped from $8-10 down to $4-5. The intense price competition continued in 1975 and drove Bowmar into bankruptcy (*Electronics*, 20 February 1975).

By 1976 calculator prices were as low as $5.99 for a calculator produced by a U.S. firm, Fantasia Calculator Corporation, and designed around a National chip (*Electronics*, 22 January 1976). As the cost of integrated-circuits fell, assembly and materials became the dominant portion of total cost, and domestic assembly became less cost effective. By 1976 the Japanese share of U.S. calculator sales was back up to as much as 30 percent. More dramatically, however, as much as 50 percent of calculators sold in the United States were manufactured offshore by U.S. firms (*Electronics*, 5 February 1976).

Watches. Integrated-circuits have successfully made their way into other consumer products. Watches are another area into which some semiconductor firms have integrated forward. Though starting slightly later than calculators, these products have also followed a pattern of rapidly decreasing prices due to intense price competition and innovation. The data we have assembled do not indicate which semiconductor firm first started producing watches, but Intel was producing electronic watches in 1973 (Lochwing 1973). By early 1974 over twenty firms marketed electronic watches, and by early 1975 forty firms marketed the product (*Electronics*, 20 February 1975). The effect of entry was, predictably, decreasing prices.

TI, Intel, and National were among the semiconductor firms manufacturing watches in 1975. National had two brand names, the new Excelar line with prices ranging from $65-85 and Timeline, for which prices were cut from $120-220 to $80-160. Fairchild entered in 1975 with watches for private-label sale (*Electronics*, 6 March 1975).

As in calculators, TI was extremely aggressive with watch prices. At the January 1976 Consumer Electronics Show in Chicago, TI displayed a watch retailing for $19.95. National, a firm also known for price aggressiveness, was caught with its prices high and lowered them in mid-show to as low as $22 (*Electronics*, 22 January 1976).

Many firms were driven from the watch business by the low prices. In 1976 and 1977 AMS, Benrus, Armin, Gillette, Gruen, Litrouix, and HMW Industries left the digital watch business (*Electronics*, 22 January 1976).

After entering in 1975 Fairchild rose to second place after TI in unit sales in two years. However, in spring 1977 TI cut the price of its bottom-line model from $20 to $10, and Fairchild did not foresee a rapid shift in preferences from light-emitting diode (LED) displays to liquid-crystal displays (LCD) (*Business Week*, 15 August 1977). Consequently, Fairchild suffered substantial losses in watches in 1977 and dropped the product line in 1978 (*Electronics*, 22 January 1976).

Conclusions. The history of the penetration of consumer products shows digital integrated-circuit manufacturers following two basic strategies. The opportunities in consumer products enticed many semiconductor firms into integrating forward, selling under their own label. Many other firms continued just as component suppliers. The other strategy that was followed was one of aggressive pricing. TI was the most successful in adapting to these new strategies, though National appears to have been successful as well. Both firms are comfortable with low-pricing strategies characteristic of the consumer-products area and in fact, TI's pricing was a significant contributor to the rapidly decreasing prices. Both TI and National were successful in the innovative-marketing step of forward integration.

Perhaps even more than other product areas, the consumer-products area was characterized by the rapid entry of many firms. Probably the major reasons for this entry were that the consumer-products area represented an incrementally innovative application of semiconductor technology, rather than major innovation, and that many of the firms formerly strong in the consumer-product areas being attacked by the MOS-LSI technology could buy circuits themselves from other integrated-circuit manufacturers and stay in the race. Relatively little technical knowledge was required to assemble the purchased integrated-circuits into finished products, and the established marketing organizations of the older consumer-products firms gave them an initial advantage over their newly forward-integrated rivals. In the end, price aggressiveness was the key determinant of success.

Conclusions

The case studies point to several general conclusions. First of all, strategic differences among firms apparently have led to vigorous competition in the semiconductor industry. In particular, the coexistence of new-product-development strategies along with second-sourcing strategies has intensified rivalry among firms. The result has been a more rapid rate of both product improvement and price reduction.

The second major conclusion is that in each of the technology races examined, the nature of the competitive process changed over time. Typically, at the beginning of a new technology race firms placed strategic emphasis on major-product innovation. In many cases this emphasis created opportunities for new innovative firms to enter the industry. Over time, as products became more standardized and sales volume increased, additional firms joined the race, often as second sources. Early participants sometimes left, either because of innovation failures or because of increasing price competition.

Finally, evidence in this chapter and chapter 3 suggests that firms can be classified into strategic groups. Firms in each group share common elements of strategy. Certain groups are more difficult to enter than others; in other words, so-called mobility barriers exist among groups. The height of these barriers is an important determinant of company profits. The relationship between mobility barriers and firm performance is considered in chapter 5.

Each of these basic conclusions is elaborated on in the next sections.

Strategic Diversity

In the semiconductor industry strategic diversity is present when some firms choose the propriety-product strategy while other firms pursue a second-sourcing strategy. The proprietary-product strategy, on the one hand, by definition leads to new-product development. The second-sourcing strategy, on the other hand, frequently leads to price competition. This competition occurs because firms that second source another firm's design generally must price more aggressively to gain sales. Firms that obtain an authorized license to second source are probably less aggressive in this regard than firms that have simply copied the original product by "reverse engineering."

The coexistence of proprietary-product and second-sourcing strategies leads to a more rapid rate of both product development and price reduction. Thus, overall performance is better than it would be if all firms emphasized proprietary-product development or if all firms emphasized aggressive price competition. By creating price competition, the second-source firm also puts pressure on the product innovator to come up with new innovations where price competition can be avoided, if only temporarily. In other words, the product-innovator firm can only maintain high profits in the long run if it develops a continual sequence of innovative improvements.

Thus, the forces of both carrot and stick serve to stimulate product-innovating firms. After introducing a successful innovation, the firm initially has the market to itself and can collect high returns. Competition soon appears in the form of second sources. Although the original firm typically authorizes another firm to serve as second source to encourage acceptance of the new product, unauthorized firms can often simply copy the design. The lack of patent enforcements and the high personnel mobility in the industry often facilitate this duplication. Price competition may then develop. Nevertheless, the initial firm may maintain a cost advantage because of its greater cumulative experience in producing the product. Eventually, as other firms gain experience, and presumably lower cost, the initial firm may be faced with decreased profit margins unless it is able to generate additional product innovations.

Examining the case histories, custom LSI apparently failed because it lacked this type of strategic diversity. In custom LSI all firms followed a proprietary-product strategy. Because no second sources evolved, strong pressure for price reduction and continual product improvement did not arise. At the same time, product and price competition were fierce in standard-product fields. Customers ultimately opted for the standard products where they could obtain the benefits of price reduction and product improvements.

Interestingly, in the semiconductor industry most firms use both proprietary-product and second-source strategies. Among the major firms Intel (which does not second source) is the only exception. National and AMD were exclusively second source when they were getting off the ground in the 1960s, but they use both strategies today. Strategic diversity exists in other industries, but individual firms typically do not follow multiple strategies. For example, Porter (1976) discusses the strategic diversity among manufacturers of branded and unbranded consumer products. The fact that in the semoconductor industry the same firms use both proprietary-product and second-sourcing strategies probably makes rivalry more intense.

Another type of strategic diversity is diversity in technological approach. Technological expertise accumulates within a firm as a result of its own technological efforts over time and its ability to hire personnel from outside. Thus, firms have different comparative advantages in technology. These differences lead firms to take alternate approaches to the same problem. In the memory field, for example, Intel developed MOS devices while Fairchild chose to perfect bipolar memories. In microprocessors, Intel led in MOS while AMD, a relatively bipolar-oriented firm has been successful in bit-slice (bipolar) technology. In general older firms have tended to modify and improve established technology, while new firms have tended to pioneer in the development of new technology. This diversity has led to a healthy exploration of alternate technological approaches. Competition has been increased, and the range of product and performance characteristics available to customers has been expanded.

Not all innovations have been successful, however. Many of the new entrants in memory technology in the late 1960s stumbled and some went bankrupt. Sylvania's TTL bipolar-logic line was technologically successful and gained customer acceptance, but Sylvania failed because of production and delivery problems. The TI and Fairchild approaches to custom LSI were unsuccessful largely because the market never developed. Such a high failure rate can be expected in an innovative industry. In part, the high-failure rate reflects the fact that firms are exploring a variety of technological options, some of which will prove to be better than others.

A related issue is the importance of timing. Being first to market with a new innovation does not guarantee success. Sylvania introduced TTL before TI but TI emerged as the winner in TTL. Advanced Memory Systems beat Intel to market with the 1K RAM, but Intel became the market leader. Motorola has been successful using the strategy of relatively late entry, thereby avoiding the frequent product "mistakes" of early entrance. In general though, early entry offers an advantage because of the important relationship between costs and the experience curve.

Competition and the Product Cycle in Integrated Circuits

The case histories in this chapter exhibit many characteristics of what has been termed the *product life cycle* (Wells 1972). Indeed, each of the individual technology races can be viewed as a separate product cycle (or part of a product cycle). The product-cycle model predicts product standardization, demand growth, and increasing capital intensity in production over time. Entry generally takes place in the beginning of the cycle, followed by a shake-out of firms as price competition becomes progressively more intense. Moreover, the theory suggests that foreign firms will enter the market at a later point than domestic firms.

The product life-cycle pattern has been observed in many industries. What appears to have been unique about semiconductors, however, is the rate at which fundamental innovations have appeared. Each of these fundamental innovations, in a sense, has renewed the product cycle. For example, as transistor technology was approaching maturity in the early 1960s, bipolar integrated-logic circuits were invented; later, as bipolar technology was maturing, advances in MOS technology enabled major breakthroughs in microprocessors and memories. The repeated reversal of the product life cycle, brought about by these and other fundamental innovations, has delayed the onset of overall maturity in the semiconductor industry. The high rate of fundamental innovation seems to have been a result of the richness of the technology base in the solid-state area. Whether the same pattern will continue in the future is an open question.

The case histories give many examples of entry at the beginning of the product cycle. In the case of semiconductor memories, for instance, established firms such as TI and Fairchild began production at the beginning of the cycle in the early 1970s; of greater interest, though, was the entry of entirely new firms into the industry at this time. Intel, Advanced Memory Systems, and other firms were launched to take advantage of technological opportunities in the memory field. Entry barriers were low because the established firms had few experience-related advantages in memories.

Other cases also illustrate the formation and entry of new firms at the beginning of a new product cycle. Zilog was founded relatively early in the cycle of microprocessors. Western Digital and MOS Technology are two of the many small MOS firms that entered early in the product cycle for electronic calculators. However, not all entry has been at the beginning of the product cycle. For example, National and AMD entered using second-source strategies comparatively late in the cycle for bipolar logic and linear circuits in the 1960s.

Some evidence exists that new firms tend to pioneer new technology whereas older firms refine existing technology. For example, Intel and Advanced Memory Systems pioneered in memories using MOS technology, while Fairchild chose to emphasize further development of bipolar memories. Tilton (1971) found even stronger evidence for this difference in emphasis in his study of semiconductor firms during the 1950s and 1960s when Fairchild and TI ranked among the new firms. This tendency contributes to strategic asymmetry, and it promotes overall technological development.

Exit has often occurred as products have approached maturity. Faced with increasing product standardization and eroding profit margins in the memory field, Intel began to withdraw from the more competitive segments in 1979 so as to concentrate its resources in more profitable fields. Firms typically withdraw because of increased price competition, as in the case of Intel, or because of declining demand resulting from product obsolescence. Motorola, however, has frequently found that remaining in these declining market segments after other competitors have left is financially rewarding.

Several of these cases show that foreign firms have lagged behind domestic firms in entering the U.S. market. Foreign manufacturers (mostly Japanese) entered the bipolar-logic and microprocessor areas at a later date than most U.S. firms. Moreover, they chose to second-source U.S. firms rather than offer their own proprietary products.

This strategic behavior by foreign firms is due to their lagging behind U.S. technology. For instance, Japanese firms were approximately one year behind U.S. integrated-circuit technology in the early 1970s (*Japan Economic Journal*, 14 December 1971). In contrast, the general belief in the early 1970s was that U.S. firms could duplicate a new product within six months of initial introduction (Finan 1975). With product cycles of only a few years, a one-year technology lag would make entry by foreign firms difficult except through second sourcing. However, many observers believe that the technology gap narrowed considerably in the middle to late 1970s, raising the possibility of foreign entry through proprietary products.

The memory field is where the technology gap has narrowed the most and where Japanese producers have been the most successful. The U.S. market share of Japanese firms began to rise rapidly in the late 1970s. By this time product standardization had reached a high level in memories, and the direction of new development had become fairly clear. This environment was well suited to the strategy of Japanese firms, which emphasized product quality and reliability in addition to competitive price.

One of the executives contacted in the interviews described how his firm's strategic priorities tend to shift over the course of the product cycle. In the case of new products, product features was ranked as the most important strategic variable, followed by technical support, delivery and marketing, reputation and past performance, and price, in that order. For older products price was ranked first, followed by delivery, features, and marketing. Thus, apparently an almost complete reversal of strategic priorities occurs as product fields grow in maturity over time.

Strategic Groups

Analysts of business decision making have traditionally studied corporate strategy in the context of the activities of a single firm. In recent years, however, several authors have integrated strategy with the study of competition. The result has been the concept of strategic groups (Newman 1978, 1979; Porter 1976, chapter 4).

Grouping firms according to the similarity of their strategies provides an analytical tool that lies between two extremes. At one extreme, each firm is unique. While there is almost always a degree of truth to this, viewing each firm as unique becomes unwieldly when a large number of important firms are under analysis. At the other extreme, much economic analysis has viewed industries as composed of relatively homogeneous firms. While analytically tractable, this extreme obscures many of the differences among firms that contribute to industry behavior.

The material in this chapter, together with that in chapter 3, can be used to define strategic groups in the digital integrated-circuits industry. Since the distinction between full-line and specialized firms captures an important part of the distinction between various product, pricing, and marketing strategies in integrated-circuits, and since this variable is also relatively easy to observe, we use it as a basis for classifying firms into strategic groups. In addition, since the distinction between firms making major breakthroughs as well as incremental innovations and firms predominantly engaged in incremental innovations is also important, and not as closely related to the other variable, we can reasonably employ this distinction also.

Firms that pursued a successful major-innovation strategy can be identified from the case histories in this chapter as well as from tables 3-1 through 3-3. TI and Fairchild were important in bipolar logic and pushed major innovations that failed in custom LSI. Intel, Mostek, and TI were involved in the major 4K- and 16K-memory innovations, while Fairchild came through with bipolar memories. Intel has been the most consistent major innovator in microprocessors. These four firms also stand out in tables 3-1 through 3-3 as the only firms, with the exception of RCA, that were involved in two or more major innovations during either the pre- or post-1968 period.

RCA is a difficult firm to categorize because it has not capitalized on its innovations to the same extent as other firms. This lack could be due to its organization, as discussed in chapter 3. It could also be due to deliberate strategy, such as emphasizing invention more than the application of the invention as an innovation. RCA's large patent holdings and royalties from patents support this possibility. Due to the difficulty of classifying RCA, we have excluded its data from the statistical analyses in chapters 3 and 5.

Using successful major innovation as a criterion for identifying strategy presents a problem, however, if some firms adopt a major-innovation strategy but fail or if some incremental strategies result in a major breakthrough because of good fortune. Some firms that were near the first in a new field might be considered to have attempted a major innovation but ended up with an incremental innovation because another firm made the major breakthrough. In the case histories firms such as Advanced Memory Systems in memories, National and Rockwell in microprocessors, and Sylvania in bipolar logic would fall in this category. We can be reasonably sure that most such attempts will have been recorded in the trade literature for those innovations that were successfully carried through as major breakthroughs by some firm. What is much more difficult to identify, however, are those attempted major innovations that never were made successfully by any firm. An example of a major innovation that never was realized is the attempt by TI (Discretionary Wiring) and Fairchild (Micromatrix) at custom LSI circuits.

About the only solution to these problems is to have information on strategy formulation prior to the events that result from implementation of the strategies. This type of information, however, is very difficult to obtain. Information about strategies in the past will inevitably be biased to some extent by what actually occurred and therefore requires the same qualifying remarks as the case-history information presented in this chapter. An appropriate characterization of this problem is that "the victors write the history." Information about current and future strategies is subject to confidentiality and even if it could be obtained, the performance resulting from current strategies could not be measured for several years—beyond the scope of most studies.

Therefore, although using successful major innovation as a criterion for identifying strategy is a problem, it is the best alternative available. We should note, furthermore, that the classification resulting from this criterion accords well with the views about innovation strategies held by people in the industry at the time these strategies were formulated, although these opinions are not unanimous (interviews with industry executives, consultants, and former industry employees).

Figure 4-1 shows firms classified according to major (plus incremental) innovation versus incremental innovation and full product-line versus specialized product variables. Several points about these groups are important. First, age of the firm is an important determinant of breadth of product line. Fairchild, TI, Motorola, and Signetics have all been in business since early in the industry and although National was revitalized in 1967, it too is older. However, age is not the sole determinant, as RCA and General Instrument have also been in business since the early 1960s and are relatively specialized.

INNOVATION STRATEGY	BREADTH OF PRODUCT LINE AND ASSOCIATED STRATEGIES	
	FULL LINE	SPECIALIZED
Major Innovations Plus Incremental	Fairchild TI	Intel Mostek RCA (?)
Predominately Incremental Innovation	Motorola National Signetics	AMD AMS AMI Intersil General Instrument Rockwell Zilog Others

Figure 4-1. Strategic Groups in Digital Integrated-Circuits

Second, mobility barriers exist among groups. Moving from specialized to full-line production is difficult. Entering a radically new product area requires hiring good new-product-area people in engineering, internal marketing, and production. For instance, a radio-frequency-transistor production worker knows little or nothing about microprocessors or other integrated-circuits, for that matter. A good team is difficult to find and put together. Assembling several such teams successfully and simultaneously is quite a feat indeed, which explains why firms take on new lines incrementally. Moving into the major-innovation group is also difficult given the risk and top-management skills involved. The easiest group to enter, incremental and specialized, is the most populated one, as would be expected.

Third, the strategies do appear to be reasonably stable. Stability of strategies facilitates the analysis of firm performance presented in chapter 5 by making it easier to compare a firm's data at different points in time. Fairchild and TI have maintained an ability to introduce major innovations through the history of the industry. This ability is particularly remarkable for Fairchild, given the exodus of people from the company and the number of top-management changes. Motorola has maintained a strategy of incremental innovation. The full-line firms have kept that position and the specialized firms stayed specialized, largely due to the mobility barriers into the full-line groups.

Note

1. Robert Noyce, "Microelectronics," *Scientific American* 237, no. 3 pp. 62-69. Reprinted with permission.

5
The Effect of Strategies on Firm and Industry Performance

The strategies identified in chapter 4 provide a useful framework for evaluating the financial performance of individual firms and the social performance, in terms of growth, prices, and benefits to customers, of the semiconductor industry as a whole. The first section of this chapter analyzes the relation of strategies to firm performance. We find that the analysis of strategic groups provides useful theoretical predictions with respect to firm performance. In particular, firms in the group that is easiest to enter, specialized product line with incremental innovation, show the worst financial performance. Major innovation and full-line production both produce financial benefits.

In the next section, the analysis turns to industry performance. The industry has performed well on most objective measures. An important reason for this good performance is the diversity of strategies found in the semiconductor industry. We also estimate the benefits to customers in a particular product area, MOS dynamic RAMs. These benefits are found to be large compared to the financial return to producers.

International performance is the subject of the third section. Past performance by U.S. semiconductor firms in world markets has generally been good. However, some observers are concerned about the future due to foreign government R&D programs, capital advantages, and barriers to trade. The results of previous chapters are used to analyze the potential for foreign competition.

Finally, the last section discusses the feedback of performance on organization and strategies. Poor firm performance has led to mergers with large equipment firms, many of which are foreign. Good industry performance, in contrast, has contributed to the withdrawal of large U.S. equipment firms. Both of these examples have important policy implications. The final section summarizes the chapter and discusses those policy implications.

Relation of Strategies to Firm Performance

Several theoretical predictions about firm performance can be drawn from the discussion of strategies and strategic groups in chapter 4. First, major product innovations should be more profitable than incremental innovations. The lead time and cost reduction accruing to a firm making a major breakthrough give considerable opportunity for profits. Second, the least-profitable strategic group should be the specialized incremental innovators because that group is the easiest to enter. This notion implies that full-line production is more profitable than specialized production within the class of incremental innovators. However, for key innovators the effect of full-line production is less clear and may be reversed. Where mobility barriers are high for other reasons, specialization may increase profits by allowing the firm to concentrate its resources in the most promising areas.

These predictions can be tested by examining data on individual integrated-circuit firms. The analysis presented here uses estimates of firms' share of bipolar digital and MOS integrated-circuit sales, growth rates of sales, and corporate profitability measures. To test for the effect of strategies on performance, a simple econometric specification will be used:

$$\text{Performance} = a + b_1 \text{ SPINN} + b_2 \text{ FLINN} + b_3 \text{ FLINC} + u$$

where

$\text{SPINN} = 1$ if the firm pursues a key-innovation strategy combined with product specialization,

$\phantom{\text{SPINN}} = 0$ otherwise,

$\text{FLINN} = 1$ if the firm pursues a key-innovation strategy combined with a full-line product strategy,

$\phantom{\text{FLINN}} = 0$ otherwise,

$\text{FLINC} = 1$ if the firm pursues an incremental-innovation strategy combined with a full-line product strategy, and

$u = $ a random-error term.

This specification tests whether performance differs among the four strategic groups discussed in chapter 4. (See the third section of chapter 4 for a full explanation of strategic groups and of the firms in those groups.) The three dummy variables represent three of the four groups. The constant term a provides an estimate of the mean performance for the fourth group, incremental innovation combined with product specialization, because firms in this group will have zero values for the three dummy variables. (A firm is placed in only one group.) The interpretation of b_1, b_2, and b_3, therefore, is the estimated difference in performance for each group compared to the specialized-incremental group. For example, the estimated performance of a firm in the SPINN group will be $a + b_1$; since the estimated performance for the specialized-incremental group is a, the difference between the two groups is represented by b_1.

The Effect of Strategies on Performance

Several different measures of performance are used including: profit-to-sales ratio and rate of return on equity averaged over a seven-year period; estimates of the firm's share of MOS, bipolar digital, and total digital sales in particular years; and growth rate of MOS, bipolar digital, and total digital sales. Table 5-1 presents definitions of the performance measures used. The main limitation of the data is that profitability for two of the

Table 5-1
Definitions of Performance Variables

Performance Variable	Definition
MOSSH73	Firm's estimated share of total U.S. MOS integrated-circuit sales in 1973.
MOSSH78	Firm's estimated share of total U.S. MOS integrated-circuit sales in 1978.
BIPSH73	Firm's estimated share of total U.S. bipolar integrated-circuit sales in 1973.
BIPSH78	Firm's estimated share of total U.S. bipolar integrated-circuit sales in 1978.
DIGSH73	Firm's estimated share of total U.S. digital integrated-circuit sales in 1973.[a]
DIGSH78	Firm's estimated share of total U.S. digital integrated-circuit sales in 1978.[a]
MOSG	The growth rate of the firm's estimated sales of MOS integrated circuits from 1973 to 1978.[b]
BIPG	The growth rate of the firm's estimated sales of bipolar integrated circuits from 1973 to 1978.[b]
DIGG	The growth rate of the firm's estimated sales of digital integrated circuits from 1973 to 1978.[a,b]
NPE	Weighted average of the ratio of net profits[c] to average equity[d] for the years 1972 to 1978.[e]
OPS	Weighted average of the ratio of operating profit[c] to sales for the years 1971 to 1978.[f]

Source: Sales figures are estimates made by Dataquest based on information from individual companies and total industry figures compiled by the Semiconductor Industry Association.

[a]The sales of digital integrated circuits equal the sum of MOS and bipolar integrated-circuit sales.

[b]Growth rate $= \left[\dfrac{1978 \text{ sales}}{1973 \text{ sales}}\right]^{1/5} - 1$

[c]Total corporate profitability including nonsemiconductor operations. TI and Motorola included. Other highly diversified firms excluded.

[d]Average equity is the average of equity at the beginning of the year and equity at year end.

[e]One observation was the weighted average of 1974-1978.

[f]One observation was the weighted average for 1973-1978.

largest firms reflects substantial nonsemiconductor operations and profit data are also not available for several subsidiaries and divisions of very large firms. However, theoretical and empirical support exists for the relationship of market share to profitability (Dalton and Levin 1977; Gale 1972). Therefore, the share of sales variables may be viewed as proxies for profitability.

Using the share of sales as a proxy for profitability will be less reliable the greater the number of specialized niches within the domain over which total sales are measured. For instance, a firm with a broad share of overall sales may have only a small share in each niche and be less profitable than firms with large shares in specialized niches but low overall shares. However, digital integrated circuits are not composed entirely of small niches. Rather, some large-volume standard products, such as RAMs, microprocessors, and logic circuits account for a large chunk of total sales, while other products occupy more specialized niches. Given this skewed distribution of sales among products, a high firm share of total sales will usually reflect high shares of some products, implying high profitability. However, some small firms may still be quite profitable if they have a high share of some small niches.

Tables 5-2 through 5-4 present the regression results. The estimated coefficients for the constant and three independent variables are presented for each of the eleven different dependent variables. As mentioned earlier, the coefficients estimate the difference in performance of each group represented by an independent variable compared to the specialized incremental group, the mean of which is estimated by the constant. For example, the coefficient on SPINN for the MOSSH78 dependent variable regression (table 5-2) means that, on average, the share of total MOS revenues in 1978 was 10.5 percentage points higher for a firm in the specialized key-innovator group than for a firm in the specialized incremental-innovator group.

As seen in table 5-2, the specialized key-innovator and both full-line groups have significantly greater shares of MOS sales in 1978 than the specialized incremental-innovator group; the specialized innovator group performed best.[1] The results for the 1973 MOS share regression indicate, however, that the shares of the three groups were higher than the specialized incremental-innovator group, but not significantly higher statistically. The results for 1973 probably differ from the 1978 results because the major innovators had not yet shown gains from MOS products such as 4K RAMs and microprocessors that were just being introduced in 1973. In addition, the improved position of the full-line incremental group between 1973 and 1978 demonstrates the strength of those firms' strategies.

In the regressions using bipolar shares shown in table 5-3, the specialized major-innovator group performs no better than the specialized incremental-

Table 5-2
Regression Results for Share and Growth Rate of MOS Integrated-Circuit Sales

Dependent Variable	Regression Coefficients (t-Statistics in parentheses)				R^2	Number of Observations
	Constant	SPINN	FLINN	FLINC		
MOSSH73	0.036 (2.43)	0.061 (1.65)	0.057 (1.56)	0.016 (0.00)	0.267	17
MOSSH78	0.019 (2.25)	0.105 (4.33)	0.069 (2.83)	0.043 (2.09)	0.599	21
MOSG	0.29 (3.35)	0.08 (0.39)	0.03 (0.14)	0.16 (0.88)	0.060	17

Source: Calculations by CRA, December 1979. Data from Dataquest, June 1979. Definitions of the variables are presented in table 5-1.

Table 5-3
Regression Results for Share and Growth Rate of Bipolar Integrated-Circuit Sales

Dependent Variable	Regression Coefficients (t-Statistics in parentheses)				R^2	Number of Observations
	Constant	SPINN	FLINN	FLINC		
BIPSH73	0.020 (1.34)	0.003 (0.068)	0.242 (7.63)	0.076 (2.77)	0.871	13
BIPSH78	0.020 (1.44)	−0.003 (−0.08)	0.203 (6.54)	0.087 (3.25)	0.827	14
BIPG	0.06 (1.57)	−0.06 (−0.55)	−0.04 (−0.44)	0.02 (0.31)	0.073	13

Source: Calculations by CRA, December 1979. Data from Dataquest, June 1979. Definitions of the variables are presented in table 5-1.

innovator group, while both full-line groups have significantly higher shares than the specialized-incremental group. These results differ from the MOS sales because the specialized major-innovator firms are new firms that have concentrated on the new MOS technology, while the older full-line firms concentrated on bipolar technology.

In the results for the digital integrated-circuit shares for 1973 shown in table 5-4, the relative novelty of MOS technology again probably explains

**Table 5-4
Regression Results for Share and Growth Rate of Digital Integrated-Circuit Sales and Corporate Profitability**

Dependent Variable	Regression Coefficients (t-Statistics in parentheses)				R^2	Number of Observations
	Constant	SPINN	FLINN	FLINC		
DIGSH73	0.017 (2.07)	0.025 (1.12)	0.185 (8.27)	0.058 (3.05)	0.817	20
DIGSH78	0.013 (2.11)	0.067 (3.41)	0.126 (6.44)	0.066 (4.00)	0.742	24
DIGG	0.12 (3.39)	0.20 (2.06)	−0.04 (−0.37)	0.07 (0.83)	0.240	20
NPE	0.04 (0.65)	0.19 (1.94)	0.11 (1.10)	0.14 (1.46)	0.434	10
OPS	0.12 (4.37)	0.12 (2.44)	0.03 (0.58)	0.04 (0.74)	0.501	10

Source: Calculations by CRA, December 1979. Data from Dataquest, June 1979. Definitions of the variables are presented in table 5-1.

the better performance of the full-line groups. However, by 1978 all three groups have significantly higher shares than the specialized incremental-innovator group.

The MOS sales-growth results are surprising. None of the groups differ significantly from the specialized incremental-innovator group. The lack of significant growth coefficients seems surprising since the 1973 MOS share coefficients are not statistically significant while the 1978 coefficients are. Several features of the data contribute to this result. Many of the specialized-incremental firms were small in 1973 and grew rapidly. However, some firms in this group had large shares in 1973 and grew slowly. This aspect of the data means that the estimated growth coefficient for the specialized-incremental group will be higher than is implied by the difference in the means (the constant terms) in the two years multiplied by total industry sales. Consequently, less variation is evident between firms in different groups in terms of growth than in terms of market share by 1978 and therefore the market share coefficient is significant while the growth coefficient is not. Note also that the estimated share for specialized-incremental firms (the constant term) actually declines from 1973 to 1978.

The bipolar sales growth also shows no difference among groups, but because little growth in bipolar sales occurred during the 1973-1978 period, one would expect such results. However, the digital sales-growth results are also surprising, for while MOS and bipolar growth show no differences among groups, the total digital sales growth is significantly higher for the specialized-innovator group.

Finally, table 5-4 also shows that with operating profits as a percent of sales as the dependent variable, the profitability of the specialized key-innovator group differs significantly from that of the specialized incremental-innovator group. The profits of the full-line groups are not different from the specialized incremental-innovator group, though these results are subject to the caveat that for each company the profitability figures are from the whole firm, rather than the semiconductor operations only. The results from the ratio of net profits to equity are similar, but suggest that the difference in profitability of the specialized key innovator is weaker, and the difference for the full-line groups stronger, although none of the coefficients is statistically significant.

The results discussed here demonstrate that the performance of firms in three of the four strategic groups is significantly better than the fourth group, specialized-incremental innovators. In addition, the data can be used to test the individual effects of the two strategy variables by applying apropriate restrictions to the estimated equations. The hypothesis that full-line production yields greater shares or profits than specialization may be tested by testing the null hypothesis that $b_1 = b_2$ and $b_3 = 0$. If these conditions were to hold then there would be no performance differences due to full-line production. Rejection of the null hypothesis means that it is not necessary to reject the alternative hypothesis that differences do exist. Similarly, the hypothesis that major innovation yields greater performance than incremental is tested by examining the null hypothesis that $b_2 = b_3$ and $b_1 = 0$.

These joint restrictions are appropriately tested by constructing a test statistic that has an F-distribution. Table 5-5 shows the F-statistics for these hypotheses for each of the share and profitability regressions. (The growth regression tests are not presented because the results were not statistically significant, as was also found in the previous discussion.) These results indicate that in four of the eight regressions the full-line variable yields significantly greater performance. Furthermore, in five of the eight regressions the major-innovation variable yields significantly greater performance. As discussed previously the profitability dependent variables suffer from diversification of some firms' operations and fewer available observations and degrees of freedom. The significant results are all found in the share regressions. If share regressions alone are considered, the full-line and major-innovation variables yield significantly greater performace in four out of six and five out of six cases, respectively.

These results confirm the predictions of the effect of strategy on firm performance. Major innovation seems to result in higher profits and sales than incremental-innovation strategies, and a full-line product strategy

Table 5-5
Test Statistics for the Hypotheses that Major-Innovation and Full-Line Product Strategies Yield Greater Performance

	F-Statistic for Null Hypothesis (degrees of freedom)	
Dependent Variable	Specialized = Full-Line ($b_1 = b_2$ and $b_3 = 0$)	Major Innovator = Incremental Innovator ($b_2 = b_3$ and $b_1 = 0$)
MOSSH73	0.003 (2,13)	2.238 (2,13)
MOSSH78	2.817 (2,17)	9.751[a] (2,17)
BIPSH73	16.004[a] (2,9)	10.606[a] (2,9)
BIPSH78	14.485[a] (2,10)	5.289[b] (2,10)
DIGSH73	19.404[a] (2,16)	11.859[a] (2,16)
DIGSH78	10.558[a] (2,20)	9.023[a] (2,20)
NPE	1.339 (2,6)	1.938 (2,6)
OPS	1.581 (2,6)	2.990 (2,6)

Source: Calculations by CRA, March 1980. Data from Dataquest, June 1979. Definitions of variables are presented in table 5-1.
[a]Reject null hypothesis with 1-percent probability of error.
[b]Reject null hypothesis with 5-percent probability of error.

seems to outperform specialization. As a result, both full-line firms and major innovators can outperform specialized-incremental innovators. The results also demonstrate the usefulness of strategic groups as an analytical tool. For instance, the poor relative performance of the specialized-incremental-innovation group can be predicted from the concept of mobility barriers into the other groups. Since mobility barriers shelter firms to some extent from competition, the least-sheltered group will have the lowest profits.

Industry Performance

Technological opportunity, firm strategies, and laissez faire government policies have combined to produce impressive industry performance. The industry has grown rapidly, prices have declined, and profits have been at levels comparable to the average for all manufacturing industries. International performance by U.S. firms, discussed at greater length in the third section, has also been exceptionally good to date. This section presents a brief overview of industry performance. Next, we discuss the importance of diversity of firm strategies for industry performance. Finally, we present an analysis of the benefits to society from a particular integrated-circuit innovation: MOS dynamic RAMs. These benefits are shown to be high compared to the financial benefits accruing to producers.

Overview of Industry Performance

The semiconductor industry has experienced rapid growth. Total integrated-circuit production by U.S.-based firms has grown from $29 million in 1960 to $3 billion in 1977 (U.S. DOC 1979, p. 39). Worldwide digital integrated-circuit sales have grown from $447 million in 1971 to $2.65 billion in 1978 (data from Dataquest). The rapid growth of the semiconductor industry can also be seen by referring to figures 2-1 through 2-3 in chapter 2.

The industry has also produced rapidly declining prices. For example, the cost per bit of random-access memory declined by an average of 35 percent per year from 1970 to 1977 (Noyce 1977, p. 67). However, this pattern may be changing. Recently the nominal prices of such products as 16K dynamic RAMs, EPROMs, and microprocessors have remained relatively stable, and the prices of some products such as bipolar memories, standard TTL circuits, and mature CMOS logic have risen (Gold 1979). However, the prices of most of these products have still declined in real terms. One reason for the absence of dramatic declining prices of some products, such as 16K dynamic RAMs, has been a temporary shortage, but suppliers' costs have also been increasing. For example, in the microprocessor area firms have incurred heavy costs for skilled labor in the development of software and peripheral products (Gold 1979). These components of total costs are not subject to cost declines through improved yields as are other costs of integrated circuits.

Profits in the semiconductor industry have been comparable to profits in other industries. Table 5-6 presents net earnings after taxes as a percentage of sales for semiconductor manufacturing and all manufacturing from 1964 to 1977. Over the period semiconductor profits averaged 4.5 percent of sales, while all manufacturing profits averaged 5.0 percent of sales. Table 5-7 presents net earnings after taxes as a percentage of equity for a sample of semiconductor firms and all manufacturing corporations from 1973 to 1977. These data confirm the pattern shown in table 5-6 for these years. With the exception of 1973 when sales increased sharply, the profitability of semiconductors has been roughly the same as that of all manufacturing. These profit rates for semiconductors are even more striking when one con-

Table 5-6
Net Earnings after Taxes as a Percent of Sales: 1964-1977

Year	Semiconductor Manufacturing	All Manufacturing Industries
1964	5.2	5.2
1965	5.9	5.6
1966	5.3	5.6
1967	3.6	5.0
1968	3.4	5.1
1969	3.2	4.8
1970	1.1	4.0
1971	2.7	4.2
1972	5.0	4.3
1973	7.4	4.7
1974	6.1	5.5
1975	3.9	4.6
1976	5.4	5.4
1977	5.1	5.3
14-year average 1964-1977	4.5	5.0

Source: All manufacturing data from Federal Trade Commission, *Quarterly Financial Report for Manufacturing Corporations, 1965-1977*; and semiconductor manufacturing data from BDBD Financial Sample, as produced in the U.S. Department of Commerce, *A Report on the U.S. Semiconductor Industry* (Washington, D.C.: U.S. Government Printing Office, September 1979), p. 57.

siders the unusually high rate of growth in that industry compared to all manufacturing and the positive effect high rates of growth normally have on profit rates.

Importance of Diversity of Strategies

One of the major conclusions emerging from the analysis of different product areas in chapter 4 is the importance of diversity of strategies in promoting innovation and industry performance. The standard product areas—bipolar logic, memories, and microprocessors—have seen the greatest diversity of strategies, such as major and incremental innovations, second sourcing, and aggressive pricing. As a result those product areas have exhibited the most dynamic performances. The product area that never lived up to expectations, custom LSI, performed poorly in large part because strategic diversity was not present. In consumer circuits, which evolved out of the custom LSI product approach, diversity returned, with aggressive pricing, second sourcing, and differences among firms in attempting to integrate forward.

Table 5-7
Net Earnings after Taxes as a Percent of Equity: 1973-1977
(*percent*)

Year	Sample of Semiconductor Firms[a]	All Manufacturing Corporations
1973	21.2	12.8
1974	16.8	14.9
1975	9.9	11.6
1976	13.9	14.0
1977	14.7	14.2

Source: For semiconductor firms calculations by CRA based on data from Dataquest, 1979.

For all manufacturing corporations data from U.S. Federal Trade Commission, *Quarterly Financial Report for Manufacturing, Mining, and Trade Corporations,* as cited in U.S. Bureau of the Census, *Statistical Abstract of the United States: 1978,* 99th edition (Washington, D.C.: U.S. Government Printing Office, 1978), p. 579.

[a]The sample includes AMD, American Microsystems, Electronic Arrays, Fairchild, Intel, Mostek, Motorola, National, Siliconix, and Texas Instruments.

In the custom LSI case history discussed in chapter 4, the established firms incorrectly perceived that customers would prefer custom products over standard products. Because of the firms' choice of custom product strategies, the product area lacked the second sourcing, ready availability, and declining prices of the product areas that included standard products. However, though the strategies of the established firms did not satisfy customer preferences, other firms such as Intel and National perceived these preferences and followed standard-product strategies that led to the dramatic performance of the standard LSI products in the 1970s. Thus, the less-than-optimal performance resulting from the relatively homogeneous strategies of the established firms in the custom LSI product area created the opportunity for success of firms with new strategies.

The following section presents a quantitative analysis of industry performance in MOS dynamic RAMs. This product area was chosen for special analysis because, although great improvements have been made in its features (for example, speed, power usage, and cost per bit), the basic logic design and application in customer equipment have remained stable. Other major product areas are more difficult to analyze because in these areas the number of circuit types and end-use applications have increased. However, these other product areas have also seen dramatic progress on particular comparable features. For instance, within half a year of introducing the first microprocessor, the 4004, Intel introduced the 8008 with ten-times-faster operation for typical instruction times (*Datamation,* December 1974,

p. 91), and in the following year the firm introduced the 8080 with twenty times the speed of the 4004 (*Fortune*, November 1975). Prices have also rapidly declined. In December 1974, the price of the 8008 was $300 (*Datamation*, December 1974, p. 91), and by the first quarter of 1977 the price of the 8080A, a product with much greater speed performance, was down to $20 (*Financial World*, 15 March 1977).

This improvement in features and decline in prices was spurred by rapid entry and the accompanying diversity of strategies. By the end of 1975, four years after the introduction of the 4004, twenty-four firms were producing or announcing mircoprocessors (Webbink 1977, p. 131).

Estimating the Benefits to Society from Integrated-Circuit Innovations: The Case of MOS Dynamic RAMs[a]

Economists often estimate in dollar terms the aggregate social benefits from technological innovations (Griliches 1958; Mansfield et al. 1977). Consumers benefit if innovations lead to price reductions or improved products. Producers receive benefits if successful innovations lead to higher profits. The total benefits to society are the sum of all such consumer and producer gains, plus any external or third-party benefits.

In this section we estimate the benefits to consumers from innovations in the design and manufacture of MOS dynamic memories. MOS memories were selected because the consumer benefits can be measured in a relatively straightforward manner. This ease of measurement is permitted by technological advances that have taken the form of cost reductions rather than fundamental design-feature changes. Improvements have been made in lower power consumption, speed of information retrieval, and ease of circuit use, but these improvements have been less significant than the rapid decline in cost per memory bit. Measuring benefits in cases in which product features have changed substantially over time is considerably more difficult.

MOS dynamic RAMs represent an important segment of total semiconductor sales. The benefits from innovations in this field have been considerable. As the estimates in this section show, advances in technology have generated a consumer surplus, which has been of the same order of magnitude as the total dollar volume of sales.

Consumer Surplus and the Welfare Gain from Innovation. The concept of *consumer surplus* provides a tool for measuring changes in social welfare. Consumer surplus is the difference between the maximum price that consumers are willing to pay for a given product and the actual price of that product. In other words, it is the amount by which the value derived from a product exceeds the purchase price.

[a]Marvin Lieberman performed the calculations and wrote this section.

Consumer surplus can be measured as the area under a demand curve and above the horizontal line denoting price, as shown in figure 5-1. Consumers purchase Q units at the market price P_1. Their total expenditure is Q times P_1 dollars and is represented by the rectangular area $0P_1AQ$. However, most consumers would be willing to pay more than the price P_1. The amount they would be willing to pay for quantity Q is measured by the entire area under the demand curve up to the point Q. (Imagine a process whereby each unit were sold to the highest bidder. The first unit would go to the consumer willing to pay the highest price, at point P_0. Additional units would be sold at successively lower prices down to price p.) In figure 5-1, this amount is equivalent to area $0P_0AQ$. The difference between what consumers would be willing to pay for Q units (area $0P_0AQ$) and what they actually pay for Q units (area $0P_1AQ$) is the consumer surplus. The consumer surplus in figure 5-1 is represented by the triangular area P_1P_0A. Consumer

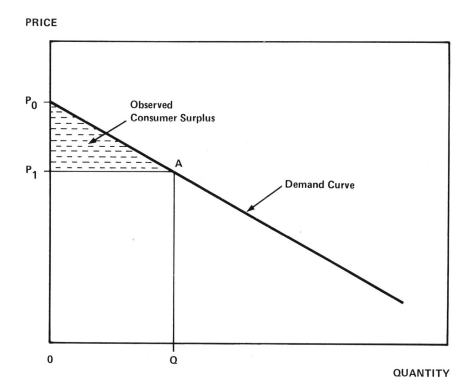

Figure 5-1. The Demand Curve and Consumer Surplus

surplus thus expressed is a measure of the social satisfaction in excess of cost, derived from having Q units available in the marketplace.

If the price of a product falls, the price line moves down and the area showing consumer surplus increases. This increase representing the gain in consumer welfare generated by the fall in price has two components. One is that consumers enjoy a greater consumer surplus for the first Q units they purchase. The second is that at the lower price they purchase more units.

As an illustration of this principle, consider the following example. Assume that a typical consumer is willing to pay up to $100 for an electronic calculator of a certain quality. If the price of the calculator is $100, the buyer will purchase the calculator, and the buyers' consumer surplus is 0. However, if the calculator costs only $75, then the buyer's consumer surplus is $25; if the calculator costs $10, then the buyer's consumer surplus is $90; and so on.

Suppose that another consumer is willing to buy a calculator at a different price, say $25. In this case, if the market price of a calculator is $10, both consumers together will enjoy $105 of consumer surplus: $90 from the first calculator ($100 minus $10) and $15 from the second ($25 minus $10). The assumption here is that the utility enjoyed by different individuals is additive.

The consumer surplus for society as a whole is simply the sum of the surpluses of all individual consumers. This sum corresponds to the area under the demand curve, since the demand curve gives the aggregate quanitity that consumers desire to purchase at each given price.

For most products the demand curve is unknown, or known only approximately. However, if the price of a product is falling rapidly over time, the demand curve and consumer surplus can be approximated using the technique indicated in figure 5-2. Assume that in year 0 the price is P_0. No units are purchased at this initial price. The next year, the price falls to P_1 and Q_1 units are purchased. The consumer surplus is then equal to the triangular area P_0P_1A. The following year, price declines to P_2 and quantity increases to Q_2. The consumer surplus is then P_0P_2B. As prices continue to fall, the demand curve is traced out. Each year the consumer surplus becomes larger. This technique assumes that all other determinants of demand except the product's price are constant. Thus, we are observing movement down a given demand curve rather than a shift in the demand curve over time. This assumption is conservative as will be shown later in this section.

Identifying the Demand Curve for MOS Dynamic RAMs. The price of a unit of MOS dynamic memory has fallen steadily at a rate of about 30 percent per year, from approximately 1.0 cent per bit in 1970 to 0.05 cent in 1979. Figure 2-5 shows that this decline can be attributed to two major fac-

The Effect of Strategies on Performance 125

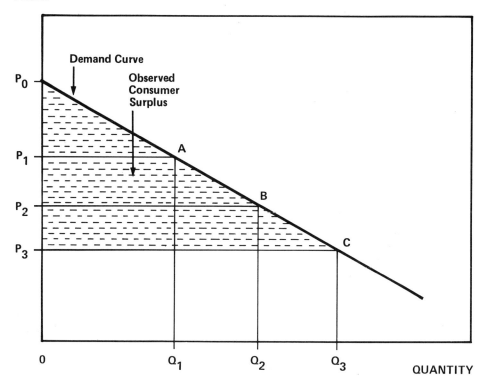

YEAR	PRICE	QUANTITY	OBSERVED CONSUMER SURPLUS
0	P_0	0	0
1	P_1	Q_1	area P_0P_1A
2	P_2	Q_2	area P_0P_2B
3	P_3	Q_3	area P_0P_3C

Figure 5-2. The Changes in Consumer Surplus that Correspond with Changes in Price

tors: (1) the experience curve and (2) increases in the number of memory bits per integrated circuit.

Over this same period, 1970 to 1979, sales grew rapidly. Dollar sales of MOS dynamic RAMs rose from approximately $10 million in 1971 to $386 million in 1979 (*Electronics*, 4 January 1973 and 3 January 1980). Since prices fell, sales in terms of memory bits rose even faster. Roughly one billion bits of MOS dynamic memory were sold in 1971, compared with nearly 800 billion bits in 1979. (These figures do not include sales of bipolar or static MOS RAMs.)

The annual price and quantity data for MOS dynamic RAMs can be used to identify the demand curve. This demand curve can then be used to estimate the consumer surplus generated by the price reductions.

Figure 5-3 plots yearly price and quantity figures, based on data from *Electronics* and from figure 2-5. Note that the graph is scaled in double-log form. Two alternative demand curves are shown. The first curve intersects the points for 1971 and 1979 but falls below the points for all intermediate years. This curve is included to provide an extremely conservative lower-bound estimate of the consumer surplus. The second curve is fitted by inspection through the points for 1973 and later years. This second curve is probably a closer approximation of the true demand curve. The period from 1970 to 1972 can be viewed as an initial "acceptance period" during which customers were still evaluating the new product, and consumption remained below the equilibrium level.

Both curves have the constant elasticity property and the mathematical form:

$$Q = \alpha P^{-\epsilon}$$

where

Q = quantity (in bits),
P = price per bit,
ϵ = the elasticity, and
α = a constant.

For any point along the curve, a given percent change in price leads to the same percent change in quantity, meaning that $d \log Q / d \log P$ = constant. For curve 1, $\alpha = 2.30 \times 10^4$ and $\epsilon = 2.28$, and for curve 2, $\alpha = 9.81 \times 10^5$ and $\epsilon = 1.79$.

Several assumptions are required for these curves to be valid representations of the actual demand curve. These assumptions are generally conservative in that they lead to a downward bias in the calculations of consumer surplus. Consequently, the true consumer surplus is probably larger than the estimates developed here.

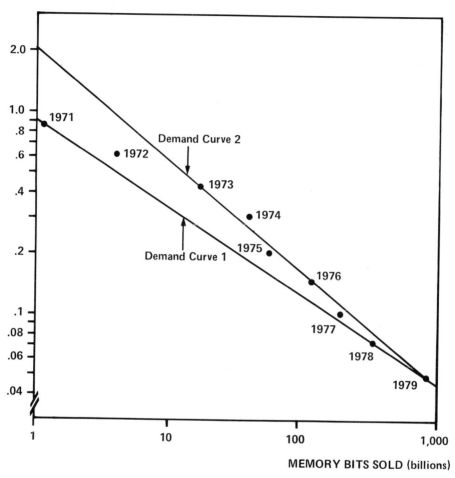

Source: Calculations by CRA using data from figure 2-5 and *Electronics*, "U.S. Market Forecasts," (January), 1973 through 1980.

Figure 5-3. Estimated Demand Curve for MOS Dynamic RAMs

One assumption is that the demand curve is stationary. However, the demand curve has, in fact, been shifting to the right each year, because the demand for semiconductor memories is derived from the overall demand for computer equipment, which has been growing. If the curve has indeed been shifting, then the actual demand curves are as depicted in figure 5-4.

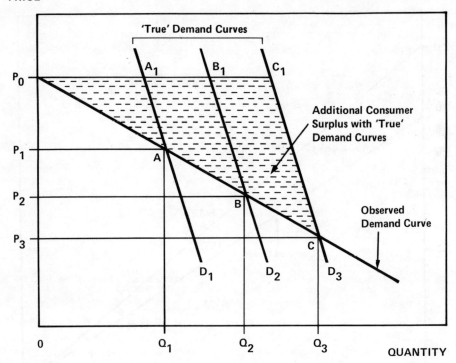

YEAR	PRICE	QUANTITY	OBSERVED CONSUMER SURPLUS	'TRUE' CONSUMER SURPLUS
0	P_0	0	0	0
1	P_1	Q_1	area P_0AP_1	area $P_0P_1AA_1$
2	P_2	Q_2	area P_0BP_2	area $P_0P_2BB_1$
3	P_3	Q_3	area P_0CP_3	area $P_0P_3CC_1$

Figure 5-4. Effect of Shifts in True Demand Curves on Calculation of Consumer Surplus

As the figure shows, use of a stationary demand curve leads to an underestimate of the consumer surplus.

The Effect of Strategies on Performance

A second assumption is that no quality improvements have been made over time. All memory bits are considered identical; units produced in 1979 are assumed to be equivalent to units produced in 1970. In fact, however, important improvements have been made in power consumption, speed of information retrieval, and ease of circuit use. The value of these improvements is not captured in the consumer surplus calculations.

A third assumption is that income effects can be neglected. Income effects arise if a change in the price of a good affects an individual's purchasing power. Since RAMs are a relatively small proportion of consumers' income, the income effects would appear small relative to other sources of error in the calculations. For instance, *Business Week* (3 December 1979, p. 68) indicates that integrated circuits accounted for less than 2 percent of the value of computers in 1970. By 1980, the figure had risen to 7 percent. Since integrated circuits other than RAMs are included, the figures place an upper bound on the proportion of consumers' (for instance, computer manufacturers') income spent on RAMs. Willig (1976, pp. 589-597) has shown that the errors due to income effects lead to relatively small errors in most consumer-surplus calculations.

A final assumption is that the demand curve is horizontal up to the point corresponding to 1970. In effect, we assume that no consumer surplus was generated until semiconductor memories became cost competitive with magnetic cores. The estimates of consumer surplus reflect gains in welfare resulting from price reductions after 1970.

Estimating the Consumer Surplus. Given the preceding assumptions, the consumer surplus in year $197X$ is given by the formula:

$$
\begin{aligned}
CS_{197X} &= \int_{P_{197X}}^{P_{1970}} Q(P)\,dP \\
&= \int_{(.7^X)\,1.25\,\times\,10^{-2}}^{1.25\,\times\,10^{-2}} \alpha P^{-\epsilon}\,dP \\
&= \left. \frac{\alpha}{1-\epsilon} P^{(1-\epsilon)} \right|_{(.7)^X\,1.25\,\times\,10^{-2}}^{1.25\,\times\,10^{-2}} \\
&= \frac{\alpha}{1-\epsilon}[1 - .7^{X(1-\epsilon)}][1.25\,\times\,10^{-2}]^{(1-\epsilon)}
\end{aligned}
$$

Table 5-8 gives annual values based on each of the two demand curves shown in figure 5-3. The estimates based on the second demand curve are

Table 5-8
Comparison of Consumer Surplus and Sales for MOS Dynamic RAMs
(*millions of dollars*)

Year	Sales	Consumer Surplus	
		Demand Curve 1	Demand Curve 2
1971	10.0	2.8	12.9
1972	24.2	7.3	30.0
1973	71.3	14.4	52.7
1974	115.8	25.5	82.8
1975	133.6	43.1	122.7
1976	162.8	70.9	175.5
1977	187.7	114.8	245.7
1978	234.8	184.1	338.6
1979	385.6	293.4	460.5

Source: Calculations by CRA.

most likely the most accurate, except for the initial years. The table reveals that in each year the consumer surplus has been of roughly the same magnitude as the dollar sales volume. In any event, the surplus has been large—probably between $300 and $500 million in 1979 alone. Moreover, this figure includes only MOS dynamic RAMs; other types of semiconductor memory would probably show similar values.

Comparison of Consumer Surplus to Producer Surplus. *Producer surplus* is the increase in profits resulting from successful innovation. In the case of MOS dynamic RAMs, the producer surplus has been small relative to the consumer surplus. As demonstrated previously, the consumer surplus has been of about the same magnitude as dollar sales. Company profits, however, are typically a much smaller fraction of sales. For example, before-tax returns for Intel, consistently the most profitable firm in the industry in recent years and a major producer of MOS memories, have ranged from 17 to 30 percent of sales between 1972 and 1978, with a weighted average of 23 percent of sales. Pretax profits for Mostek, another major producer of MOS memories, ranged from −5 percent to 30 percent of sales during the same period, with a weighted average of 11 percent of sales (financial figures from Dataquest).

Some of these profits represent the cost of capital invested in facilities to produce the innovation. The remainder would be producer surplus. Since consumer surplus is five to ten times greater than total profits, apparently consumer surplus is at least that much greater than producer surplus. Producer surplus is relatively small, in part because of the vigorous competition in this industry. Most "innovation rents" have been quickly competed away, going to consumers in the form of lower prices rather than to producers in the form of higher profits.

International Performance

Since its inception the U.S. semiconductor industry has been the leader in worldwide semiconductor sales. U.S.-based firms have traditionally accounted for nearly all U.S. sales, around 10 percent of Japanese sales, and approximately 50 percent of European sales. The U.S. share is higher in more-advanced products such as integrated circuits and lower in older products such as discrete semiconductor devices. Table 5-9 shows estimates of the shares of U.S.-based firms in Japanese, European, and worldwide discrete and integrated-circuit shipments.

Table 5-9
Estimated Share of U.S.-Based Companies in Total World Shipments of Discrete and Integrated-Circuit Semiconductors: 1974-1977
(*percent*)

	1974	1975	1976	1977
Europe				
Discrete	40.0	38.9	38.0	36.7
Integrated circuits	70.1	67.1	69.9	67.7
Total	51.6	50.4	51.9	51.1
Japan				
Discrete	4.2	3.3	5.9	4.6
Integrated circuits	22.2	18.3	20.2	16.6
Total	11.4	9.9	12.7	10.7
Rest of world				
Discrete	42.5	43.6	52.0	44.1
Integrated circuits	26.6	34.3	41.4	39.7
Total	35.7	39.4	47.0	41.9
World[a]				
Discrete	51.2	50.8	52.2	48.8
Integrated circuits	72.5	69.2	71.2	69.4
Total	61.1	59.6	62.0	60.1

Source: Calculated by CRA using tables in ICEC, *Status 1978: A Report on the Integrated Circuit Industry*, 1978, pp. 1-8, 1-9. Reprinted with permission.
Note: Shipments of U.S.-based companies used to calculate percentages include shipments of U.S. foreign subsidiaries.
[a]Including the United States.

The share of U.S.-based companies in total Japanese semiconductor shipments has been fairly small and constant at about 10 percent, with the U.S share of integrated-circuit shipments at roughly 20 percent of Japanese shipments (Tilton 1971, p. 144). Total integrated-circuit imports equal 25 percent of Japanese integrated-circuit production (or 20 percent of apparent consumption). The U.S. share of these imports declined from nearly 100 percent in the late 1960s to a steady 60 percent since 1972. Correspondingly, Germany's share of Japanese integrated-circuit imports has risen to 10 percent in 1976 from less than 1 percent a few years ago (Pacific Projects, Ltd. 1977).

The position of U.S. firms in Japan has been limited both through selective buying of U.S. products and through restrictions on U.S. direct investment in the Japanese semiconductor industry. In general, the Japanese in the past have imported only those semiconductor products that are not available domestically, effectively limiting the growth of U.S. imports to the most advanced integrated circuits (Noyce 1979). However, access to the Japanese market has reportedly improved since the Tokyo round of trade negotiations in 1979.

Except for TI, U.S. direct investment in Japan's semiconductor industry is limited to 50 percent ownership in joint ventures. TI gained its 50-percent joint venture with Sony in 1967 only after threatening court action for Japanese infringement of TI's integrated-circuit patents (Tilton 1971). Even after gaining 100 percent ownership in 1972, TI is still required to disclose certain patents and restrict its output (*Japan Economic Journal*, 4 January 1972).

The European semiconductor industry is dominated by two types of firms: old, established European receiving-tube firms and subsidiaries of U.S. semiconductor firms. Very little entry and exit of European firms has occurred since 1960. Most of the receiving-tube firms that began semiconductor production in the 1950s still exist today. Beginning with TI in 1957, virtually all the new entry into the European semiconductor industry has been from U.S. foreign subsidiaries.

Two important international aspects of the semiconductor industry should be kept in mind in interpreting the statistics. First, many U.S.-based firms rely on offshore assembly of semiconductor devices. The most advanced stages of production, R&D and wafer fabrication, are usually done domestically. The fabricated chips are then sent to offshore sites for bonding and assembly into packages. The assembled device is then sent back to the United States or to other destinations.

A second important point is that recently a number of foreign firms have acquired control of or a minority interest in some U.S.-based semiconductor firms. The motives and consequences underlying this trend are discussed in chapters 2 and 6.

Despite the historically strong U.S. international position in semiconductors, much concern currently exists about the future. Some policy makers and industry executives view Japanese firms in particular, but also European firms, as threats to the position of U.S. firms. Several issues underlie this concern. First, foreign governments have funded large R&D programs aimed at giving their firms an edge in future semiconductor technology. Second, Japanese firms have had a sheltered domestic market, as discussed previously. Third, foreign firms, particularly Japanese, have access to long-term debt capital on more favorable terms than U.S. firms.

The analysis of strategic groups and the determinants of innovation in chapters 3 and 4 provide important insights into these concerns over future international performance. The foreign R&D efforts, such as Japan's VLSI (very large-scale integration) program, can be viewed as attempts to propel some foreign firms into the key-innovation strategic groups. However, the results in chapter 3, suggest that this view should not be accepted blindly. Two of the most important determinants of key innovations, top-management involvement and above-average risk taking, are promoted by the laissez faire U.S. system. Another important characteristic, the balance between organizational flexibility and management control, has been refined by a number of strong U.S. semiconductor firms. The difficulties that large U.S. electronics firms such as RCA, Sylvania, and Philco-Ford had with this characteristic can be expected also to face foreign semiconductor firms, since most of them are more similar in organization to the diversified U.S. electronics firms than to the semiconductor specialists. Thus, the critical characteristics facilitating entry into the key-innovation groups have been present for U.S. firms but not for foreign firms.

The difficulty of foreign firms entering the major-innovation groups, however, must be qualified in at least two respects. First, as U.S. semiconductor firms have grown they have become more like other large firms in some aspects of organizational structure. If a firm as large as TI can maintain an ability to introduce major innovations, other large foreign (and domestic) firms will possibly be able to adopt similar management and organizational characteristics for their semiconductor operations. Second, as the industry matures technologically the opportunities for major innovation may decrease considerably. If fewer major breakthroughs were to be made, the fact that foreign firms are at a disadvantage with respect to making major innovations would make less difference.

The one characteristic with which foreign firms do have an edge is capital availability and, through government funding, R&D spending. However, this characteristic was found in chapter 3 to be more closely related to incremental innovation than major innovation. In addition, we find in chapter 6 that few major innovations have resulted from government-funded programs in the United States. Thus, the main advantages foreign firms have with respect to innovation would appear to promote incremental rather than major innovations.

What can be expected, then, is that foreign firms will become formidable competitors in the incremental-innovation groups. This expectation is reinforced by the fact that most foreign semiconductor firms, with their in-house relationship to diversified electronics applications, would qualify as full-line rather than specialized semiconductor producers. As discussed previously, firms in the full-line-incremental group are usually more successful than the firms in the specialized-incremental group since the latter is the easier group to enter.

A final piece of evidence reinforces these conclusions. Japanese firms have made their greatest penetration into U.S. integrated-circuit sales in the 16K-RAM product area. In 1979 Japanese firms were reported to account for 40 percent of 16K RAMs in the United States (*Business Week*, 3 December 1979, p. 86). RAMs, however, are the most standardized LSI product. Some U.S. firms such as Intel have moved out of RAM production to free resources for other more-advanced areas. Thus, the Japanese gains to date are exactly where one would expect—in a relatively older, more-standardized product area where incremental innovation and the production and marketing advantages of a full-line firm will have the greatest impact.

The threat of foreign competition through incremental innovation should not be taken lightly, however. As discussed in previous chapters much of the innovative effort by firms in the major-innovation groups goes into incremental advances, as well, of course, as the effort by firms in the incremental groups. In addition, the U.S. firms in the incremental groups hold a substantial share of digital integrated-circuit sales. Currently the firms in the full-line-incremental group account for approximately 26 percent of U.S.-based firms' sales (counting Signetics as a U.S.-based firms, which is debatable given the firm's ownership by Philips). The firms in the specialized-incremental group account for approximately 34 percent of U.S. firms' sales.

Although analogies with other industries can be overdrawn, we should note that similar groups in the U.S color-television-receiver industry held similar shares in the late 1960s prior to major gains by Japanese firms (CRA 1979). Much of the Japanese innovative effort in color televisions would be classified as incremental, although one firm, Sony, would be classified as a specialized major innovator. Two major U.S. innovators, RCA and Zenith, maintained a combined market share of roughly 45 percent despite Japanese gains in U.S. sales. However, smaller, more-specialized and incremental U.S. firms suffered substantial loss of market share to Japanese firms.

The Feedback of Performance on Organization and Strategy

Financial performance has a feedback effect on the strategies and organization of the firm, for performance is a major criterion with which the firm evaluates its strategies. Firm performance also affects firm strategies and organization through its effect on capital availability and cost. Examples of the feedback effect are found in several semiconductor firms.

Fairchild was a very successful firm in the semiconductor industry in the early 1960s but suffered losses in the late 1960s in the semiconductor division as well as in other divisions. Fairchild attempted to achieve profitability by making a number of organizational changes. In 1967 and 1968 Fairchild went through three executive officers before hiring C. Lester Hogan for Motorola (*Electronics*, 30 September 1968, pp. 119-122). Hogan in turn brought seven persons from Motorola to head the company (*Electronics*, 2 September 1968).

Fairchild also provides an example of strategy changes in response to performance. Prior to 1968 Fairchild refused to produce any product as a second source (Golding 1971, p. 163). Its sole emphasis was on proprietary designs. However, after performance deteriorated and new management took over, Fairchild became more like other full-line firms by producing both proprietary and second-source designs.

Firms that fail with proprietary designs often second source another firm's product (interview with industry executive). Fairchild had such an experience in the microprocessor area. As mentioned in chapter 4, Fairchild achieved success with its F-8 two-chip microprocessor but was beaten by Mostek in introducing a one-chip version and consequently second sourced the Mostek device.

Other examples of the feedback effect are found in the strategies of AMD and National. AMD entered the semiconductor industry as a small firm. With little capital the firm pursued a second-source product strategy. However, AMD's long-run strategy was to manufacture proprietary designs (interview with industry executive). Profits from the second-source products provided the capital necessary for the development of proprietary designs, and AMD shifted in the mid-1970s to a strategy of proprietary designs in some product areas.

The National experience is similar. Because of growth of the company, its risk taking has changed over time. In its early years the company only pursued projects that would pay back within six months, but today it accepts projects with payback periods of several years (interview with industry executive).

Takeovers and mergers are forms of organizational change. Small, specialized firms that have not performed well are particularly subject to this kind of change, for they find it difficult to raise capital other than from established firms, and their depressed value makes them attractive to other firms. A reflection of this phenomenon is the fact that only seven of the thirty-six semiconductor firms started since 1966 remain independent (*Business Week*, 3 December 1979, p. 66). GMe, Advanced Memory Systems, AMI, and Electronic Arrays are examples of small firms that have suffered poor financial performance and have subsequently merged with other firms. GMe, an early MOS pioneer, was unable to exploit the new technology and was bought by Philco-Ford (Kraus 1973, pp. 50-52). Advanced Memory Systems, which offered the first 1K RAM, could not keep up with other firms and merged with Intersil in 1976 (*Business Week*, 22 November 1976). Northern Telecom Ltd. of Canada later bought 26 percent of Intersil (*Business Week*, 15 October 1979). AMI, another MOS pioneer that has not done well financially, sold 25 percent of its stock to Robert Bosch of West Germany in 1978, and Bosch later sold half of its holdings to Borg-Warner. Electronic Arrays, which also had financial difficulties, was bought up in 1978 by Nippon, a Japanese firm (*Business Week*, 3 December 1979).

Poor firm financial performance is, however, only one of a number of causes of takeover, and larger, more-successful firms such as Fairchild and Mostek have also been subjects of takeover (*Electronics*, 10 May 1979, pp. 46, 138). But though Fairchild and Mostek are not failures, both have had problems. Fairchild's growth, particularly in MOS technology, has not kept up with the industry. Though Mostek is currently doing well, it has experienced periods of losses and low profitability (*Business Week*, 10 September 1979).

Industry performance can also have a feedback effect on firms' behavior. The success of a firm or a group of firms in new-product areas induces the entry of other firms as, for example, in the case of microprocessors. Poor performance of a group of firms may also affect the behavior of firms outside the group. In the case of custom LSI the relatively poor performance of the established firms (that is, in light of the opportunities that LSI offered) induced new firms such as Intel and National to enter with standard LSI products. When these new firms achieved success, many of the established firms switched their strategies to standard products. Another example of the feedback of poor industry performance might be the Japanese inroads into 16K dynamic RAMs in the United States in 1979. When U.S. firms were unable or unwilling to expand capacity to meet demand, Japanese firms expanded capacity to increase their share of U.S. sales (*Business Week*, 3 September 1979).

A final example of the feedback of industry performance on firm behavior may be the withdrawal of many large firms such as General Electric, Westinghouse, and Sylvania from integrated circuits and the lack of entry into merchant sales by other large firms, many of which produce integrated circuits internally. Aside from the organizational disadvantages of these firms with respect to innovation as discussed in chapter 3, their lack of participation may reflect the generally good performance of the merchant semiconductor industry. The industry has performed well dynamically in terms of technological change and statically in terms of price competition, while profits have hovered around the average for all manufacturing. Large electronics firms could rationally view the integrated-circuit industry as efficiently supplying their component needs. Their decision not to participate reflects the fact that they can earn greater returns in other lines of business than in semiconductors (see tables 5-6 and 5-7).

Summary and Policy Implications

The semiconductor industry in general, and digital integrated circuits in particular, has performed extremely well. Innovation has brought major improvements in product features and reductions in cost. The industry has grown rapidly and prices have in general declined at a steady rate.

This chapter showed that firm strategies have a significant effect on firm performance. Of the strategic groups identified in chapter 4 the one that was the easiest to enter, specialized products with incremental innovation, exhibited the worst financial performance. The major-innovation groups showed the superior financial performance expected from that strategy. In addition, the full-line product strategy, with accompanying marketing and pricing strategies, produced financial benefits.

Good industry performance was found to result from having a diversity of strategies including major innovation, incremental innovation, second sourcing, and aggressive pricing. The poorest-performing product area, custom LSI, lacked this diversity of strategies.

Consumers' benefits from innovation and competition in integrated circuits were estimated for the case of MOS dynamic RAMs. Despite several conservative assumptions the benefits to consumers were found to be high compared to the financial benefits achieved by producers. By our measure, the total consumers' benefits were approximately equal to total sales of the product.

International performance by U.S.-based semiconductor firms has historically been good. U.S. firms have accounted for nearly 100 percent of U.S. sales, 50 percent of European sales, and 10 percent of Japanese sales. Current government and industry concern over foreign competition was analyzed using the results from chapters 3, 4 and 5. We found reasons for skepticism with respect to the possibility of foreign entry into the major-innovation groups. However, foreign firms can be expected to be formidable competitors in the incremental-innovation groups due to foreign capital availability, government R&D efforts and advantages associated with full product lines.

Finally, examples of the feedback on strategies and organization of firm financial performance and industry performance were discussed. Poor financial performance has led to some changes in strategies but more commonly to changes in organization. In particular, purchase of all or part of a firm by larger firms, including many foreign firms, has been common. Industry performance has affected strategies and caused the withdrawal of some large firms.

The results in this chapter have several important policy implications. First, despite the high social benefits compared to private benefits, exemplified in the analysis of MOS dynamic RAMs in the second section, strong financial incentives still exist for major innovation, as shown in the analysis of firm performance in the first section. The relatively laissez faire U.S. system has encouraged major innovation, primarily through making it financially attractive for individuals and keeping legal barriers to entry and personnel mobility low. These policies should be continued or revived to the extent that regressive changes have been made in recent years.

Second, the analysis of international performance suggests that policy should be concerned with strengthening incremental innovation by U.S. firms. The advantages that foreign firms have in terms of capital and government R&D funding should be carefully assessed since, as discussed in chapter 3, capital availability and R&D spending have the greatest effect on incremental innovation. To promote continued good international performance by U.S. firms, federal policy should concern itself with putting U.S. firms on equal footing with foreign firms in terms of capital availability and incentives and support for R&D.

A third set of conclusions relates to merger policy. The fourth section presented examples where lackluster firm financial performance led to mergers with larger firms. This information suggests that mergers, at least from the viewpoint of the acquired semiconductor firm, are a symptom of the problem of rising capital requirements and tight capital availability. From that viewpoint, policy should be concerned not with the mergers, but with the tight capital availability that promotes the mergers. By providing access to capital, the mergers can be expected to strengthen incremental innovation, with consequent implications for competition. These and other issues with respect to merger policy are discussed at greater length in chapter 6.

Note

1. The t-statistics shown in parentheses allow tests of statistical significance. A t-statistic of approximately 2.1, for instance, indicates that the estimated coefficient is significantly different from zero at the 95-percent confidence level (two-tailed test) when there are 17 degrees of freedom as in the MOSSH78 regression.

6
Analysis of Government Policies in the Semiconductor Industry

The six government policies that have had a significant impact on the semiconductor industry were indentified in table 1-1 and briefly discussed in chapter 1. In this chapter, a framework for policy analysis is applied to these policies in the context of their impact on innovation in the semiconductor industry. [Chapters 2 and 6 of CRA (1980) give a complete description of this framework.] First, we examine the differential impact of several policies among firms pursuing different strategies. Since the impact of policy on strategies is of key importance, we highlight this impact first. Then we discuss the individual policies with an in-depth look at their general impact on the development of the industry. Finally, we take a look at the interdependent effects of various policies.

The application of our framework for policy analysis to the semiconductor industry is intended to clarify the ways in which government policies have affected innovation in the industry. The analysis is also intended to illustrate how the framework for policy analysis operates and how policy makers may utilize it in addressing current policy questions. Examples are given indicating the various effects among dimensions of the framework, drawing upon evidence and links already presented in earlier chapters of this case study.

Each of the six policies identified in chapter 1 are discussed individually in the context of three distinct time periods. As discussed in chapter 1, the early years (1950s and early 1960s) of the industry were marked by strong government support, the middle years (mid- and late-1960s) were a period of laissez faire, and the recent years (1970s) were a period of increased tension between the industry and government. Along with this historical perspective current policy issues are also presented, drawing upon our interviews with industry executives and industry consultants for pertinent information. Next, the question of policy interaction and interdependency is addressed. We emphasize identification and analysis of a few key interdependencies rather than trying to isolate all instances of interaction between policies. Finally, we discuss the usefulness of our framework for analyzing and assessing government policies and their impact on innovation.

Differential Impact of Policies among Strategic Groups

In our analysis of the effects of government policies on the semiconductor industry two key points have emerged. First, many policies affect firm strategies through their impact on entry. Policies such as procurement, antitrust, and those affecting venture-capital availability allowed easy entry and made it difficult for established firms to maintain a sheltered position. Barriers to entry and mobility barriers were low and these policies worked to keep them low. Further analyses of this point are presented in the sections on individual policies.

The second point involves the different effects of a policy on various strategic groups. Understanding this differential impact is important since policy makers often expect a policy to impact an industry and the firms in the industry in a relatively uniform way. However, by examining the impact across differing strategy groups, it becomes clear that the policy impact differs, and thus the effect on competitive behavior and performance may not be the originally anticipated effect. In this section, a brief analysis of several policies is presented demonstrating this differential impact.

Government R&D funding and tax policies affecting capital availability are two policies with particularly strong effects on certain strategic groups. Incremental innovators are most affected in their strategies by capital availability and R&D spending (see chapter 3). Therefore, policies such as R&D funding and the general level of taxes can have a strong influence on the performance of incremental innovators. As discussed in chapter 5, the effect of incremental innovation on international performance must be analyzed carefully. Federal policies can have a strong impact on firms pursuing incremental-innovation strategies and policy makers concerned with R&D funding, tax policy, and international-trade policy should be cognizant of the effects of these policies on the competitive capabilities of U.S. firms in relation to foreign firms.

Our interviews also gave us information regarding different reactions among firms to the Japanese threat and policies designed to combat that threat. Executives from one firm that tends to be an incremental innovator were clearly concerned about the Japanese threat and felt that strict policy rules should be adopted against Japanese competition in the United States. In contrast, executives at a firm characterized by a major-innovation strategy were more concerned about general economic policies to keep the U.S. economy growing than about Japanese competition (interview with industry executives). One explanation for this difference is that Japanese firms are more likely to be incremental innovators than major innovators and the incremental innovator can expect more head-to-head competition with them than can the major innovator. The Japanese can be expected to have a more-difficult time competing with the major innovator's products and therefore are likely to target their sales at the incremental innovator's products. Those semiconductor firms that can expect to bear the brunt of Japanese competition are the strongest advocates of a restrictive policy.

Another policy concern of recent times is the different reaction of various semiconductor firms to the Defense Department's VHSIC program. This program is aimed at developing a high-speed-integration capability for use in advanced weapons systems. It is both a research and a procurement program. This six-year development program, which could get underway in 1980, will cost an estimated $200 million and is expected to involve all three military branches as well as several semiconductor firms (*Electronics*, 14 September 1978). The proposed program will cover research in the following areas: advanced lithography, fabrication-technology devices, systems demonstrations, and design-software-testing-architecture concepts. The goals are to increase throughput by a factor of 100 and actual delivery of military-specification devices (Weisburg 1978, p. 25). The projected military applications include radar, weapons-control systems, and electronic warfare (*Electronics*, 14 September 1978).

Its advocates hope that the VHSIC program will help keep U.S. firms abreast of the latest technology and have considerable commercial spillover (interview with industry executive). The program is also seen as a direct response to the Japanese and European government efforts in the VLSI area (Gutmanis 1979, p. 134). Commercial spillover is expected in areas such as satellite communications, weather forecasting, and search systems (Weisburg 1978).

Nevertheless, several problems have arisen in attempting to get this program started. The first problem has been the industry's response. Several firms, and in particular Intel, have voiced strong opposition to the program (*Electronics*, 28 September 1978). Intel believes that the VHSIC program will divert resources from promising commercial areas to military areas with little commercial relevance. Military technology is said to lag behind commercial technology, and the time span from design to deployment may take up to ten years, in which time technology in other areas changes drastically (interview with industry executive). Other firms believe the program will not provide the basic research needed to support the industry several years hence. In addition, some firms contend that government needs can be met with current technology and off-the-shelf devices (*Business Week*, 27 November 1978).

Intel believes this massive government effort is ill-timed and aimed in the wrong direction. According to Intel's Robert Noyce, the program will divert resources from promising commercial areas to military areas with little spillover potential. Military technology tends to lag behind commercial technology, and commercial applications are emphasized by firms whose primary strategy is to remain on the competitive edge. However, other firms have generally welcomed this new government interest and infusion of capital into the industry. They believe this program may stimulate innovation and growth in an important area of technology similar to the procurement and R&D programs of the the 1950s and 1960s. One problem, however, is that government programs such as VHSIC and others tend to be engineer intensive, while design and process engineers are presently the industry's scarce resource. Firms have expressed skepticism about their ability to commit large numbers of these engineers to the VHSIC program given that they are already working on other in-house, commercially oriented research programs (*Electronic News*, 17 December 1979). These firms assume that the payoffs from commercial research programs are likely to be higher and more immediate than for research done in conjunction with the government.

Using the strategic groups identified in chapter 4, the differential impact of this policy among strategy groups may be visualized. On the one hand, firms that pursue full-line incremental-innovation strategies seem to react much more favorably to government research funding than do most specialized or full-line key innovators. One explanation for this reaction may be that since capital availability for R&D funding is of more critical importance to the incremental innovator, these firms tend to be very receptive to any type of government R&D funding. However, full-line key innovators and specialized key innovators (especially firms with little interest in the product areas of the government's research) tend to be less receptive, and some have been critical of the VHSIC program.

Recent mergers in the semiconductor industry also may be analyzed by examining the differential impacts among strategic groups. The merger wave has raised concerns about the impact on innovation and competition and whether foreign acquirers are gaining access to U.S. technology at bargain prices. One potential issue is that a merger could divert resources from merchant semiconductor production to captive production and thereby lessen competition in merchant sales. (We refer here to "actual" competition. The captive producer would still be a potential competitor.) However, none of the acquirers of large semiconductor firms are such large end users that such a problem would arise, and the semiconductor firms that have been acquired solely for captive production are small enough that the impact on competition would be negligible.

A second potential concern over acquisitions is that large firms have had a poor record of keeping up with innovation and competition in the semiconductor industry. This might seem to be a particular problem, since two recently acquired firms, Fairchild and Mostek, have included major innovation in their strategies. However, the past behavior in this industry suggests that if major opportunities continue to exist in the future, other firms will seize them if the current leaders do not. Furthermore, people have learned from experience the consequences of organizational inflexibility. Mostek and its acquirer, United Technologies, for instance, plan to follow a hands-off policy with respect to the semiconductor operations (*Business Week*, 15 October 1979).

A third concern that has been raised over mergers is that mergers involving foreign acquirers might give foreign firms access to U.S. technology. However, except for Fairchild and Signetics, all of the acquisitions by foreign firms have been in the group of firms that our analysis in chapters 4 and 5 has labeled specialized-incremental innovators. These firms stand to benefit from foreign ownership through access to foreign firms' technology and capital. Signetics, owned by Philips, also stands to benefit from the considerable technological resources of Philips. [See table 3-1 through 3-6 and the statements of Signetics President Charles Harwood (*Business Week*, 3 December 1979)]. Schlumberger, which acquired Fairchild, is not an electronics firm and can therefore be expected to capitalize on Fairchild's technology primarily through Fairchild's own operations, a result no different than if Fairchild were owned by a U.S. firm. Thus, to date foreign acquistions do not appear to threaten the technological preeminence of U.S. firms. The firms that have been acquired stand to benefit as much from foreign technology as the foreign firms do from the acquired firms' technology.

Although patents have not been of great importance to the industry per se, patent-licensing practices may give some indication about a firm's strategies. For example, strategies may be identified based on the various motives for patent licensing. For firms following key-innovation strategies, liberal licensing arrangements were important as a means of remaining at the forefront of the industry. They licensed primarily for the access to information that new patents of other firms could give them for *future* innovations. Incremental innovators relied on liberal licensing arrangements as a means of imitating and second sourcing new products in a short time. Thus, firms in this strategic group licensed for the informational value new patents could give them for *current* innovation. Yet, since all firms had an incentive (albeit a slightly different one) to license, this became an industrywide practice.

Procurement Policy

Government procurement of semiconductor devices was extremely important during the early years of the industry. Although its relavite importance to the industry has subsequently declined, we must understand its impact and role in stimulating the early growth of the industry. Table 6-1 presents data for government purchases of semiconductors over time. The

Table 6-1
U.S. Shipments of Semiconductors for Defense Consumption: 1963-1977

Year	Total Semiconductor Shipments (millions of dollars)	Defense Shipments (millions of dollars)	Defense as Percent of Total
1963	600	213	36
1964	—	—	—
1965	879	194	22
1966	1,055	254	24
1967	1,074	297	28
1968	1,189	274	23
1969	1,457	247	17
1970	1,337	275	21
1971	1,519	193	13
1972	1,912	228	12
1973	3,458	201	6
1974	3,916	344	9
1975	3,001	239	8
1976	4,968	480	10
1977	4,583	536	12

Source: U.S. Department of Commerce, Bureau of Census, *Current Industrial Reports*, Series MA-175, "Shipments of Defense-Oriented Industries."

defense portion of total shipments declined from a level near 40 percent in the early 1960s to around 10 percent in the 1970s. Prior to 1963 the defense portion of total shipments ranged between 36 and 39 percent except for 1959 when the defense portion rose to 45 percent (Schnee 1978).

The Department of Defense (DOD) and NASA were major purchasers of semiconductors during the 1950s and early 1960s. These agencies made a strong commitment to the use of semiconductor devices, believing they could be more cost effective and yield better performance in military and space programs than could alternative technologies (Braun and MacDonald 1978, pp. 80-82). In addition to accounting for a large portion of total demand, the government was willing to pay substantially higher prices, only part of which reflected the higher cost of military-specification quality assurance. Government demand was also relatively stable and secure (Golding 1971, p. 336). As a result, procurement policy had a direct impact on the risks firms faced, particularly market risk. By providing an assured market for these new devices for a relatively long period of time, costs could be reduced and attention focused on developing additional applications of the devices. The government absorbed the risks of new production by providing an assured outlet at predetermined prices, thus allowing firms to plan their long-term production. This stability allowed firms to improve their yields more rapidly, since they did not have to be concerned with fluctuations in the volume and pace of manufacturing (Golding 1971, pp. 341-342).

In addition to reducing market risk, procurement contracts accelerated the process of innovation. For example, one early contract to TI for integrated circuits required design, fabrication, and delivery of eighteen different devices in a six-month period (Golding 1971, p. 335). The time span from invention to commercialization was cut radically (Gutmanis 1979, p. 125). Although no key innovations may be directly linked specifically to procurement policy, procurement contracts apparently accelerated the development of new products, particularly the integrated circuit. In other words, while procurement policy may not have led to semiconductor innovations that otherwise would not have occurred, it did advance the timing of some innovations. This timing was important especially in the 1950s and early 1960s, since it enabled U.S. firms to gain the innovative and competitive edge in the world market.

The combination of the rapid development of new devices and the assured demand provided by the government induced many firms to pursue innovative strategies and new-market opportunities. As military production proceeded, learning economics were realized, costs fell, and firms were able to price products more cheaply, enabling them to penetrate consumer and industrial markets. Prices for integrated circuits decreased rapidly; in 1962, integrated circuits were about $50 for one device, but by 1968 their price had fallen to $2.33 (Tilton 1971, p. 90).

Military emphasis on performance criteria in their procurement contracts and the flexibility of the contracting officers in overseeing contracts were also important in enabling the industry to grow and experiment in new areas. Performance specifications such as high reliability, low power consumption, and/or durability rather than design criteria allowed firms more diverse and creative ways of achieving the specifications and resulted in more innovation. Contracting officers were also more flexible in the 1950s and early 1960s and willing to negotiate changes in contract terms if they could be convinced of the merits of such a change. For example, one firm persuaded the contracting officer to permit it to change from developing an alloy-switching transistor to a germanium-mesa transistor for the Minuteman Missile project. The germanium-mesa transistor proved to be a success that had considerable spillover into the ability to make silicon devices (interview with industry executive).

The environment in which firms competed was indirectly affected by procurement policy in the early years of the industry. Procurement contracts tended to encourage new firms to enter the market, thereby contributing to a highly competitive market environment (Tilton 1971, p. 91). New firms often targeted their strategies towards the military, at least initially, since they knew the demand was strong and premium prices would be paid for their devices (Schnee 1978). Also, these younger firms, eager to capture a foothold in the market, actively sought military business and were flexible in tailoring their designs to government needs.

Several firms started out with procurement contracts both in the 1950s and 1960s. Transitron, which introduced the gold-bonded diode in 1953, is the most often-cited case. This diode was contracted to the military because it was capable of handling much higher voltages than other diodes (Braun and MacDonald 1978, p. 71). Throughout the 1950s and early 1960s, this product remained Transitron's major revenue producer and was sold primarily to the military. Transitron emphasized efficient production and largely ignored R&D, yet by the late 1950s became the second-largest semiconductor producer and perhaps the most profitable (Braun and MacDonald 1978, p. 71). Transitron eventually left the industry in the early 1970s, due in part to inflexibility in its strategies.

TI became the other leading firm in the late 1950s, to a large extent because of lucrative government contracts. The Air Force was particularly interested in TI's development of the silicon transistor and the integrated circuit and gave TI the first Minuteman II contract in 1962 (Golding 1971, p. 336).

Newer firms tended to be favored by the government in early contracts, in part due to the flexibility of the firm to meet government specifications. New firms may have viewed the rewards from innovations associated with government contracts as opening up a new market rather than displacing existing products as could be the case for established firms. Incentives for individuals might also be greater in small new firms than in established firms because the initial sales of new products affect sales and profits to a greater extent the smaller the firm (Utterback and Murray 1977, p. 19).

Procurement contracts also had impacts on firms' organization and the need to train engineers to deal with contracting officers. Since government demands are often highly specialized, personnel specifically designated to handle government work had to be trained. Production people within the industry disliked government business due to the problems of producing high-reliability devices for the military. However, marketing people liked military business because of its lucrative potential, thus leading to internal division concerning a firm's strategy with respect to the military demand.

Spillover from government work into the commercial sector occurred in several ways. For example, the production experience TI gained in manufacturing integrated-circuits for the Minuteman program benefited its commercial-production process, and other firms also gained knowledge and expertise in this way (Golding 1971, p. 342). Also, because of the tough performance standards imposed by the government, only a small proportion of the devices manufactured would meet these criteria. The remaining devices could often be sold at a lower but still profitable price in less-exacting civilian applications because of the high prices received on sales to the government.

As the industry moved into the 1960s, military demand declined in importance relative to total semiconductor demand (Braun and MacDonald 1978, p. 41). During the early 1960s, defense fluctuated and continued to grow but at a much slower rate than total demand (see table 6-1). Nevertheless, the invention of the integrated circuit and its rapid diffusion were greatly aided by the government, especially the NASA Apollo program and the Air Force Minuteman II project. Again, the importance lay in the timing; integrated-circuit technology and its use were rapidly diffused and readily accepted, in part because of the government's acceptance of this new technology.

TI and Westinghouse received the first Minuteman II contracts in 1962, and in 1963 other contracts went to TI, RCA, and Westinghouse. In late 1963 it was estimated that the Minuteman II contracts accounted for 60 percent of the total integrated-circuit production to that date. The bulk of the integrated-circuits produced for the Apollo project came from Fairchild; in 1963 the Apollo contract led the Minuteman II contracts in terms of number of devices delivered (Golding 1971, p. 336). The contract called for delivery of 200,000 devices from Fairchild. These two programs dominated integrated-circuit sales into 1964 and several new-weapons projects were initiated compelling not only TI, Fairchild, and Westinghouse, but also other competitors such as Motorola and Signetics, to enter into military sales.

While total semiconductor demand mushroomed due to new applications such as in computer systems, most observers and industry officials overestimated the importance that military sales would have in the late 1960s (Braun and MacDonald 1978, p. 93). Total production for government use increased 23 percent between 1964 and 1971, whereas total semiconductor production rose 140 percent during that time.

Table 2-1 indicates this shift in importance of the military segment of the industry. Competitive behavior became more important, since several firms were competing for the government business that was representing a smaller share of total output. Nevertheless, several firms got their start in the 1960s by doing government business.

Signetics was founded in 1961 and became a major military supplier in the mid-1960s, specializing in digital bipolar devices. Advanced Micro Devices (AMD) also built to military specifications in its early years as a strategy to enable it to gain a toe hold in the market (Gutmanis 1979, p. 128).

MOS technology, although less useful to the military than bipolar, was also supported by the government. NASA was particularly interested in MOS and increasing its reliability.

By the late 1960s, military demand continued to decline relative to total demand, and procurement contracting officers became more cost conscious. NASA's Apollo project was operational and no longer needed large quantities of new semiconductors. In addition, military spending dropped in the late 1960s and early 1970s, and these agencies were forced to become more cost conscious in their procurement contracts.

Stiff competition also forced prices for standard devices down and firms became reluctant to fill custom orders. Custom prices remained high, but due to the drop in standard-device prices, demand for custom devices dropped (see chapter 4 for a further explanation of the failure of the custom LSI market). Thus, firms became less willing to fill military orders, since most high-reliability military work tended to be custom in the sense that special tests had to be performed. Also, certain firms spent a large share of their research resources on military work, and thereby may have allowed competitors to gain the edge in other product areas (interview with industry executives).

In the mid-1970s military (and NASA) demand continued to decline relative to total semiconductor demand. Budget cuts at both NASA and DOD and the de-emphasis of certain programs (such as the space shuttle), requiring large amounts of electronic components, also contributed to this decline. Procurement contracts continued to stress cost specifications. Also, many firms were reluctant to accept contracts because DOD programs were perceived to be mission-oriented and designed to solve only one piece of the total problem.

During the late 1960s and early 1970s a shake-out of several firms that had historically been military suppliers occurred. Sylvania, Transitron, and Westinghouse had traditionally relied on military demand for a large segment of their semiconductor business. Chapter 3 describes the management, organization, and strategic reasons for the failure of these firms, but another contributing factor was the decline in military demand and the failure of the companies to penetrate commercial demand. Some firms were not able to cope with the new environment of the more cost-conscious military purchasers and intense price competition in sales to civilian users (OECD 1968, p. 74).

Recently, government purchasers have relied primarily on established firms as suppliers, decreasing the opportunities for new entry or successful spinoffs. In 1978, government sales of integrated circuits comprised only 10 percent of total sales (*Status* 1979, p. 4-3). Since the mid-1960s cost competition and product standardization related to the integrated circuit have reinforced these trends. Not only had government policy changed, but also the industry had matured. The industry's performance and growth were no longer crucially dependent upon the level of government demand.

Government Funding of R&D

Government funding of R&D for semiconductors originated in the DOD and NASA as a means of developing new technology to meet the nation's military and space needs. These government agencies became convinced at an early stage of the importance of semiconductor electronics to the military and space programs and therefore pushed for the development of more-reliable, faster, and smaller circuits. The Air Force was an important source of R&D funds through its research programs for missiles and air defense. Between 1959 and 1964, Air Force integrated-circuit R&D funding accounted for about 55 percent of total government R&D related to integrated circuits. Government R&D funding represented approximately 25 percent of total R&D on integrated circuits in 1960 (U.S. DOC 1961, p. 13). This "technology-push" factor had an indirect impact on the industry by emphasizing the importance of research in this area and thus stimulating other firms to conduct extensive research activities (Kleiman 1966, p. 68).

R&D funding by the public sector proceeds in a manner similar to procurement. Basically, the government is buying a product as it does in procurement, but that product in this case is the research (and/or development) of a new product or process. During the 1950s the majority of R&D funds went to the established firms that had good reputations for research. However, by the early 1960s many of the newer, highly innovative companies were being given lucrative research contracts. RCA, Motorola, and TI were three of the major recipients of R&D funds in the early 1960s.

The effect of R&D funding on firms' strategies impacted their innovative behavior in two ways. R&D funding by the government functioned to emphasize the urgency of the situation, thereby inducing greater R&D investments by private firms (Kleiman 1966, p. 68). This "urgency effect" cut development lead time since firms saw the lucrative benefits of these new devices. The early funding by the Army Signal Corps of semiconductor-production facilities to Western Electric, General Electric, Raytheon, RCA, and Sylvania provided additional stimulus for development of electronic devices. Throughout the middle and late 1950s, government funding covered engineering design and development and private firms funded production facilities, adding to total industry capacity (Tilton 1971, p. 92).

A second important impact of government R&D funding on innovative behavior was the reduction in the technical risk in developing new devices. Technical risk was reduced in two ways: (1) government R&D funding increased the private return-risk ratio firms faced in developing new products since they needed to invest less in R&D; and (2) firms' perception of risk may have been altered, especially in the 1950s and 1960s since this large-scale government action may have increased their confidence that a technical breakthrough was possible. A firm's financial structure was also affected, since the receipt of government R&D funds gives the firm more money to commit to a new product or design (Golding 1971, p. 337). As chapter 3 indicates, both capital availability and risk taking have been important determinants of innovative behavior for most firms, and R&D funding by the government aided in both respects.

Entry conditions were also affected in the early years of the industry by the pattern of government R&D funding. In contrast to procurement policy that favored new firms, the majority of R&D funds went to established firms (Tilton 1971, p. 95). In 1959, Western Electric and eight other receiving-tube firms (old firms) were receiving 78 percent of government R&D funds. Established firms may have been favored in this case because these firms were more interested in doing broad, general research and had a strong record of past performance and success in conducting research efforts. Market structure may have been affected, although by the late 1950s and early 1960s, two new firms, TI and Motorola, began receiving extensive government R&D funding.

The generation of basic knowledge and skills gained from government-funded R&D created important spillover effects in the semiconductor industry. During the 1950s, the spillover from government-funded R&D on the commercial sector was primarily indirect, however, Golding (1971 p. 342) states that "devices designed specifically for military/space application cannot normally be sold in the nongovernment sector without some modification." The "real" spillover was indirect: "The firm benefited through the extensive transfer of production experience gained during the course of the program to the manufacture of commercial devices."

An example of this direct spillover is Fairchild's micrologic circuits being used both in the NASA Apollo project and for industrial and commerical computers (Golding 1971, p. 342). Firms have the incentive to choose military R&D contracts that have maximum spillover potential. Often firms engage in cost-sharing contracts such as the Motorola-Air Force contract in 1961. This contract was designed to develop linear integrated circuits for the Air Force, but also supplemented Motorola's extensive research on using integrated circuits in communications and electronics equipment (interview with industry executive).

Universities also received extensive government funding of R&D. The Office of Naval Research supported significant research at various universities during the 1950s. Several universities were doing leading-edge work, including research into gallium arsenide at the University of Illinois. Purdue, MIT, and the University of Pennsylvania were other schools doing top research.

Analysis of Government Policies

Research efforts into increased reliability were also an important stimulus to the industry with obvious spillover potential. Both NASA and the Air Force funded research programs to improve reliability and failure analysis (interview with industry executive). The NASA Microelectronic Reliability program set criteria for the acceptance of all semiconductor devices and also required the inspection of all production facilities under contract to NASA.

Nevertheless, some conflicts developed in the early years of the industry about how and where research efforts should be centered. For example, the military itself did not achieve a consensus about how to proceed with research into miniaturizing components. Project Tinkertoy, conducted by the National Bureau of Standards in the late 1940s and early 1950s was an effort backed by the Navy and designed to increase miniaturization of electronic components and to automate their manufacture. This project ultimately failed and funding was cut off in 1953. In 1957 the Army Signal Corps revived interest in miniaturization with its Micromodule Project. By 1963, $26 million had been spent on this program, with much of the funds going to RCA. The project ended in failure, short of its goal of producing 1 million units a year (Braun and MacDonald 1978, p. 67). Finally, the Air Force financed a molecular-electronics program aimed at miniaturization. Some $2 million was given to Westinghouse to direct research in this area (Braun and MacDonald 1978, p. 100).

Although this multipronged approach to miniaturization eventually failed and the development of the integrated circuit circumvented this activity, some observers maintain that nonetheless the undertaking was valuable, because it stimulated the thinking of creative people about miniaturization and possible ways to do it. Many pioneering steps are necessary, even though they never get into production. False starts are often necessary to get working on the problem and enable the firm to recognize the solution once it has been found (interview with industry executives).

Government funding supported research into MOS technology during the 1960s. The military was particularly interested in the research being conducted at RCA and funded some of it (interview with industry executives). CMOS was of particular interest to the military because of its high-performance characteristics such as low power dissipation, high noise immunity, and fast switching speed, and in 1963 the Air Force funded research at RCA into CMOS technology, leading to its eventual development in 1967. Silicon-on-Sapphire research was also funded by the government as an extension of the research into CMOS and like CMOS was undertaken primarily by RCA.

Performance was affected in the 1950s and early 1960s as a result of the government R&D programs. Two obvious impacts were the reduction in lead time and the improved reliability of the devices. The government role as R&D sponsor enabled firms to commit further resources to R&D.

Observers have noted that government R&D spending is most effective when it complements the work already undertaken by a firm (Kleiman 1966, p. 70). This was, apparently the case in the industry in several instances. Bell Laboratories work after the invention of the transistor and Motorola's and TI's work developing germanium-mesa transistors and integrated circuits are examples.

The indirect effects of R&D funding, such as the generation of basic knowledge and the speed up of development and production, appear to have had the strongest impacts on performance. The guaranteed market and performance specifications had important ramifications for the development of the industry. Also, the cooperation among industry firms, universities, and the government surrounding early R&D efforts increased the impetus for technological change in the 1950s and early 1960s (Tanaka 1979, p. 8). The Navy-IBM-Harvard Mark I computer project and the NASA space race represented large-scale joint ventures that significantly advanced technological levels (Hogan 1979, p. 9).

During the 1960s the government role as R&D funder began to diminish: total funding tended to fluctuate over time, depending on the relevant programs under way, but funding did decline in the late 1960s. Table 6-2 presents data regarding government R&D spending in semiconductors from 1955 through 1961. Although exact data are not available for later years, R&D funding seems to have continued to increase after 1961, particularly due to the NASA efforts. However, spending peaked in the middle to late 1960s and began to drop, as did procurement funding due to budget cutbacks and lack of interest.

Research programs became more goal oriented in the 1960s, with flexibility for research becoming restricted (interview with industry executives). Also, research focused less on basic than applied (and development) re-

Table 6-2
U.S. Government Semiconductor R&D Expenditures: 1955-1961
(*millions of dollars*)

Year	Total R&D
1955	3.2
1956	4.1
1957	3.8
1958	4.0
1959	6.3
1960	6.8
1961	11.0

Source: U.S. Department of Commerce, BDSA, *Semiconductors: U.S. Production and Trade* (Washington, D.C.: U.S. Government Printing Office, 1961).

search, and funding of university research dropped considerably. In addition, federal R&D programs were not as well structured, and research efforts tended to be scattered too widely to be effective (interview with industry executives). According to an industry executive, much of the money went to "second-echelon" firms—that is, firms with little primary interest in semiconductors. Spillover into the commercial sector declined in the late 1960s due to the fragmentation of DOD and its mission-oriented slant.

Another problem with government-funded R&D has been the government's insistence on claiming title to patents generated from government-funded R&D. NASA began this trend and now additional government agencies (for example, the Department of Energy) claim title to such patents and license them freely with no royalty payments. This practice may discourage firms from undertaking government-funded research efforts since they cannot derive any significant advantage from the technical know-how generated by such a project. Firms also are reluctant to consult and interact with government engineers because they fear competitors may gain access to this information through the Freedom of Information Act.

During the 1970s government support of R&D subsided significantly, with important effects on the industry (interview with industry executives and venture capitalists). Basic research in electronics is not being performed now as extensively as in the 1950s. Universities, with a few exceptions (such as Cornell, Arizona, Stanford, MIT, and Berkeley), are not doing the leading-edge work they once were and are receiving less funding from the government. Internal funding generated by universities is not sufficient to keep firms on the technological forefront (Hogan 1979, p. 8). Companies have to fund the bulk of their R&D; with the rising costs of R&D this becomes more difficult.

Currently, estimates show that government R&D funding comprises about 5 percent of the total R&D outlays for semiconductors (Interview with industry executive). Also, none of the leading firms currently does major amounts of government-funded research (interview with industry executive). Nevertheless, many executives and (policy makers) favor increased government R&D spending. Direct support of R&D is important because many foreign governments (Europe and Japan) are supporting large semiconductor research programs in contrast to the United States (interview with industry executive). Total U.S. Government R&D outlays at DOD and NASA are expected to increase about 9 percent in 1980 (Gutmanis 1979, p. 131).

Antitrust Policy

Although U.S. antitrust policy has had only a limited effect upon the innovative record of the semiconductor industry, some evidence indicates that it has had an impact upon the industry's environment and structure as well as upon the strategies of individual firms. Antitrust policies have had impacts on the semiconductor industry at two distinct times in its development. The first period was in the middle and late 1950s. The 1956 Consent Decree negotiated between AT&T and the Justice Department had important consequences for the direction of structural and competitive growth in the semiconductor industry. The second period was the middle and late 1970s, when some industry executives expressed concerns about the heightened competitiveness of Japanese firms and the ability of U.S. firms to respond within the limits of the U.S. antitrust laws.

In 1956 AT&T settled a seven-year antitrust suit with the Justice Department concerning a patent-pooling agreement among General Electric, RCA, Westinghouse, and AT&T. According to the Justice Department, the agreement gave the Bell System a near monopoly over U.S. telecommunications equipment. Western Electric was allowed to remain part of AT&T but was enjoined from selling any output, including semiconductors, to commercial customers. Western Electric was allowed to continue production for internal Bell usage and for military and space applications (Tilton 1971, p. 50; Golding 1971, p. 135). In addition, the company was forced to license all its existing patents royalty-free to any domestic firm and guarantee licensing at reasonable royalty charges on all future patents (Tilton 1971, p. 76; Kraus 1973, p. 28). Previously, Bell made available its patents on the transistor to any company for royalty of 5 percent of sales (until 1953 when the rate was reduced to 2 percent) (Webbink 1977, p. 98). Since Bell's major activities were centered in telecommunications and it was dependent on innovations in many areas of electronics and discrete components for its success, it was willing to exchange information about new technologies. Other firms followed suit, entering into cross-licensing agreements as a means of staying abreast of new developments in the various areas of electronic components (Tilton 1971, p. 74).

During the antitrust suit AT&T refrained from any measure that might indicate an attempt to dominate the fledgling semiconductor market (Tilton 1971, p. 76). Technological advances were openly publicized and Bell Laboratories held several conferences and seminars to inform industrial representatives about their innovative progress (Kraus 1973, p. 26; Tilton 1971, p. 75; Braun and MacDonald 1978, p. 54). A liberal patent-licensing policy was pursued.

Fear of similar antitrust problems may have encouraged other large receiving-tube firms to follow AT&T's liberal licensing policy. This easy access to important patents in the early and mid-1950s allowed many new firms to enter the market (Tilton 1971, p. 77), which increased competition and shifted the emphasis from large, multidivision firms to smaller, highly specialized firms. Both market structure and competitive behavior were affected. The environment became more competitive with new entry, and the liberal licensing practices rapidly diffused the stock of basic knowledge. Competitive behavior was influenced as some firms altered their strategies. Antitrust policy apparently shifted to some extent the strategy of older firms away from competing for leading semiconductor market positions and towards manufacturing for internal use and generating revenues from innovative efforts in their other markets (for example, Bell Laboratories in telecommunications). The fast-growing new firms relied heavily upon second sourcing and often gained a competitive advantage over the older firms (Braun and MacDonald 1978, pp. 122, 137). Second sourcing sometimes provided the resource base and cash flow that could be used to develop proprietary devices. Some of the older firms like Sylvania, Sperry Rand, Westinghouse, and General Electric eventually had to drop out of the so-called merchant market altogether, although they continued to manufacture for themselves (Braun and MacDonald 1978, pp. 136-137).

Currently, antitrust policy is of some concern to industry executives because of the increase in international competition. Some industry people are upset that the antitrust laws are not evenly applied to all firms competing for sales in the United States, particularly to what they allege are unfair business practices of foreign firms in the United States. They also contend that restrictions on the size of U.S. firms hinder their ability to compete on a worldwide basis.

The two issues of greatest significance are charges that Japanese firms are "dumping" semiconductors in the United States and that Japan unfairly protects its domestic sales from foreign competition. A brief analysis of these charges and their relevance to U.S. antitrust policy is important in understanding why these issues are of such current controversy.

Dumping is defined as selling in foreign markets at lower prices than at home and is essentially price discrimination (Kindleberger 1968, p. 155). It does not necessarily involve selling below cost. In simplest terms the dumping firms must have market power at home, which permits them to enjoy a higher price there than they receive in export markets. Economic theory indicates that a firm with market power that faces different demand elasticities in different markets will achieve greater profits by charging a higher price in the market with the less-elastic demand (the protected home market) and a lower price in the markets where demand is more price elastic (the foreign markets where it faces greater competition), rather than the same price in both markets.

The second issue—that the Japanese market is essentially closed to foreign firms—is tied to the first issue. Since access by foreign firms to the Japanese market is restricted, Japanese firms have a substantial degree of market power in the home market. Thus, these firms are able to charge higher prices for semiconductors in Japan and lower prices aboard, particularly in the United States. The inability of most U.S. firms to penetrate the Japanese market, coupled with the strong competitive pressure exerted by the Japanese in the United States, has upset many U.S. semiconductor executives (interview with industry executive).

Recent evidence suggests that the Japanese may be more willing to open up their market. In light of new international trade rules (adopted in 1970) and increased U.S. pressure on the Japanese government, foreign entry into Japan may become easier in the future. Alternatively, market-share restrictions on Japanese firms in the United States and greater flexibility in the enforcement of our antitrust laws have been suggested to counteract the Japanese encroachment (*EDP Weekly*, 13 August 1979).

Some industry executives believe that the antitrust laws restrict U.S. firms from growing to the size necessary to complete with the large, highly integrated Japanese firms. However, if economies of scale in production are not high (see chapter 2), firm size may not be important for achieving competitive success.

Other potential advantages of size are marketing economies, discussed in chapter 4 with respect to full-line production, and capital availability, discussed in chapters 3 and 5. An important disadvantage of size is the difficulty in achieving a proper balance between organizational flexibility and management control as discussed in chapter 3.

The Japanese also have been accused of unfairly enforcing their patent holdings (by aggressively applying their own patents and infringing those of U.S. firms) in the United States to gain market power (interview with industry executive). Large U.S. semiconductor firms allegedly fear antitrust action by the Department of Justice if they try to enforce their patent holdings as a competitive tactic. Thus, many executives believe Japanese firms are "allowed" to compete unfairly with U.S. firms.

Trade Policies

As the volume of world trade in semiconductors and digital integrated-circuits has increased in recent years, the trade policies of the United States and other nations have taken on added importance. U.S. trade policies affect foreign firms; also, U.S. firms have been deeply affected by foreign government policies. The primary trade issues in the United States today concern access by U.S. firms to the Japanese market and subsidies by foreign governments to their semiconductor industries. These issues are of deep concern both to policy makers and industry officials. A brief history on the pattern of trade will be presented before dealing with these current policy issues.

Prior to 1965, U.S. trade in semiconductor devices was relatively

Table 6-3
World Apparent Consumption of Semiconductors
(*percent of total noncommunist-world consumption*)

Area	Years					
	1956	*1960*	*1965*	*1970*	*1974*	*1978*[a]
United States	80	76	66	53	48	46
Western Europe	8	12	18	22	25	23
Asia	10	10	14	22	24	28
Rest of non-communist world	2	2	2	3	3	3

Source: U.S. Department of Commerce, *Report on the U.S. Semiconductor Industry* (Washington, D.C.: U.S. Government Printing Office, September 1979), p. 88.
[a]Semiconductor Industry Association estimates.

minor; export sales in no year exceeded 5 percent of total factory sales (Finan 1975, p. 90). Consumption of semiconductors abroad was minimal and imports into the United States were quite low. This low amount of trade was due to the differing uses of semiconductors in the United States and abroad and the small size of foreign demand. In the United States, semiconductors were used in military and industrial applications, whereas abroad they were used primarily in computers and consumer products. Even though exports to Europe and Japan remained low, they did account for a significant share of total consumption in those countries. Table 6-3 presents data showing world consumption of semiconductors for selected years from 1956 to 1978. Also, foreign semiconductor firms were few and new, and tended to lag one to four years behind the United States in technology (Finan 1975, p. 13).

The motivating factor behind foreign imports of U.S.-produced semiconductors in the 1960s was access to new technology. U.S. trade policy imposed virtually no restrictions on exports to other free-world countries. This policy impacted the environment directly by diffusing information regarding new *product* developments to foreign competitors. Foreign firms were able to study and copy these new innovations, thereby allowing them to develop new products more rapidly than if export restrictions had prevented this diffusion. Of course, new-process innovations were much more difficult to obtain and copy until offshore assembly and direct foreign investment by U.S. firms occurred. Even with offshore assembly, R&D and the technology-intensive processes (for example, wafer fabrication) continue to be done domestically.

After 1965, a shift in the pattern of trade occurred and the total volume

Table 6-4
U.S. International Trade of Semiconductors: 1967-1977
(*millions of dollars*)

	1967	1968	1969	1970	1971	1972	1973	1974	1975	1976	1977[a]
Exports[b]	152	204	346	417	371	470	848	1,247	1,053	1,400	1,503
Imports[b]	43	72	104	157	179	330	619	961	803	1,107	1,352
Trade balance	+109	+132	+242	+260	+192	+140	+229	+286	+250	+293	+151

Source: U.S. Department of Commerce, *Report on the U.S. Semiconductor Industry* (Washington, D.C.: U.S. Government Printing Office, September 1979), p. 59.
[a]Preliminary data.
[b]Not adjusted for Tariff Items 806.30 and 807. The foreign-trade data include semiconductor subassemblies and parts of semiconductors exported for assembly or further processing in overseas plants that are then returned to the United States as imports under TSUSA items 806.30 and 807.00.

Analysis of Government Policies

of trade increased significantly. Table 6-4 indicates the pattern of U.S. trade and demonstrates the rapid increase in trade volume. U.S. firms began to set up offshore assembly plants and invest directly abroad in complete production facilities. In part, this setup was due to domestic competitive pressure to reduce costs, especially in the labor-intensive assembly phase of production (U.S. DOC Report 1979, p. 69). Note that this move offshore also forced the Japanese firms to go offshore (Chang 1972). Foreign consumption of semiconductors rose steadily throughout the late 1960s and 1970s. Prices for semiconductors abroad began to drop significantly, making foreign tariffs a problem for U.S. firms exporting into foreign countries while trying to remain competitive in their prices (Finan 1975, p. 94). Also, end uses for the devices became more similar in the United States and abroad, thereby increasing the number of foreign countries U.S. firms might penetrate with similar devices. Finally U.S. trade policy remained open, while several foreign governments began protecting their semiconductor industries with various types of trade barriers.

High tariffs and high nontariff barriers (relative to the United States) spurred U.S. firms to set up production facilities abroad (Finan 1975, p. 84). These facilities are integrated and include a total production facility rather than simply final assembly and wire bonding. In both Germany and France, unofficial quotas on the importation of semiconductors and tariffs of up to 17 percent existed; in Japan certain quotas remain in existence. However, the quota on importation of integrated-circuits containing more than 200 elements was lifted in Japan in 1974 (Finan 1975, p. 95). These restrictions provided incentives for U.S. firms to locate production facilities in these foreign nations.

US. trade policy throughout the 1970s has remained liberal, with low tariffs on imports and various measures designed to stimulate exports. U.S. imports of semiconductors are taxed at a 7-percent rate, whereas tariffs on semiconductors imported into other countries range from 10 to 20 percent (Tanaka 1979, p. 17). Thus, foreign firms face an incentive to export their devices to the United States rather than to other foreign nations that have higher tariffs. This incentive is one reason U.S. imports have risen significantly in recent years. Domestic demand, particularly in recent years, has outgrown domestic production capacity and the United States remains the largest semiconductor market in the world.

Several policies and policy issues have been important or potentially important to the industry. One such policy is the special tariff provisions of Items 806.30 and 807.00 of the U.S. Tariff Schedule. These tariff provisions tax only the value added abroad upon reimportation into the United States with an ad valorem rate of 5 to 8 percent (Finan 1975, p. 69). This policy has aided U.S. firms in their strategy to open offshore assembly plants in less-developed countries. Offshore assembly began in the late 1960s and

Table 6-5
Adjusted U.S. Trade Data for Semiconductors: 1970-1975
(thousands of dollars)

Trade	1970	1971	1972	1973	1974	1975	Average Annual Growth Rate (1970-1975)
Exports (less U.S. content[a] of 806.30/807.00)	417,021 −78,409 338,612	307,528 −81,255 289,273	469,644 −127,346 342,298	848,454 −185,637 662,817	1,247,498 −310,359 937,139	1,053,495 −291,718 761,777	17.6
Imports (less U.S. content[a] of 806.30/807.00)	157,464 −78,409 79,055	179,092 −81,255 97,837	330,277 −127,346 202,931	618,613 −185,637 432,976	961,338 −310,359 650,976	802,687 −291,718 510,969	45.3
Balance	259,557	191,436	139,367	229,841	286,168	250,808	

Source: U.S. Department of Commerce, *Report on the U.S. Semiconductor Industry* (Washington, D.C.: U.S. Government Printing Office, September 1979), p. 68.

[a]U.S. content denotes value of device upon leaving the United States prior to final assembly. Thus values are the same for both imports and exports.

Analysis of Government Policies

early 1970s with plants set up in countries such as Malaysia, Singapore, Korea, Taiwan, and Mexico to utilize low labor costs for the labor-intensive wire-bonding and other assembly steps (see chapter 2). The ability to reimport assembled products at a low tariff increased the cost savings from setting up offshore assembly plants. One source estimates that if the special tariff items were repealed, additional costs of offshore assembly would amount to 1 to 2 percent of the total cost (U.S. DOC 1979, p. 74). To show the importance of these trends, in 1977 $1.12 billion worth of semiconductors were imported under Items 806.30 and 807.00 (U.S. DOC 1979, p. 58).

Some of the increase in foreign trade of semiconductors merely reflects rising imports and exports under these special tariff provisions (Webbink 1977, p. 120). As more offshore plants were built, the volume of international trade naturally increased. Between 1960 and 1965, 13 offshore plants were built; by 1970, 71 were built; and by 1974, 135 had been constructed (U.S. DOC 1979, p. 84). Table 6-5 includes U.S. semiconductor exports and imports with adjustments made for all items exported and reimported under Items 806.30 and 807.00.

Current policy issues with respect to foreign trade revolve around the rapid increase in imports of semiconductors into the United States and the strength of competition from foreign firms, especially the Japanese. For example, the issue of dumping was dealt with in the previous section. Another issue of great concern involves foreign government subsidies to their electronics firms. This support allegedly gives foreign firms unfair advantages in competing and developing new technologies. The Japanese government, for example, is backing a research effort into VLSI amounting to $250 million (*Business Week*, 3 Decmeber 1979; the Japanese government is also alleged to support its semiconductor firms financially by "guaranteeing" their solvency (*Forbes*, 26 November 1979). Japanese electronics firms tend to be highly levered and therefore can earn higher returns on equity with relatively lower operating rates and prices. Also, the financial backing of the Japanese banks allows these firms to pursue aggressive capacity-expansion plans without the large earnings base required of U.S. firms. In Europe, government support also has risen. The French government is spending $140 million on five companies and joint ventures in the area of VLSI research. Great Britain and West Germany are also spending sizable sums to develop better microelectronics capabilities. In Great Britain the funding totals about $110 million and the West German government has allocated $150 million over several years (*Business Week*, 3 December 1979, p. 80; Scace 1979).

Nevertheless, these massive spending efforts must be viewed in context of the research efforts by U.S. firms. In 1978, U.S. semiconductor firms spent over $400 million in R&D; also the DOD will commit more than $150 million over a period of several years to develop VHSIC technology. Thus, U.S. firms are able to spend large sums toward developing new devices and new technology. A recent report notes that despite the increased activities of foreign firms and governments, the U.S. share of the worldwide market is not expected to drop below 50 percent in the 1980s (*Business Week*, 3 December 1979).

It is also important to note that the more intense competition between U.S. and Japanese firms currently centers around one product: the 16K RAM. This device, however, is one of the most-advanced and widely demanded products in the semiconductor industry today. In the three years since the device was introduced, the Japanese have made rapid progress, capturing over 40 percent of worldwide sales in 1979 (Business Week, 3 December 1979). The real fear expressed by U.S. executives is that the Japanese will parley this success into several others and eventually dominate all semiconductor sales (*Forbes,* 26 November 1979).

Tax Policies

In addition to trade policies, another set of government policies affecting the semiconductor industry has drawn attention in recent years—that is, tax policies and various tax measures, which allegedly have diminished the incentives of firms to innovate.

Tax policies have become important to the semiconductor industry only in recent years. In the early years of the industry, tax policies had little direct impact on firms. This trend continued throughout the 1960s until 1969, when the tax rate on capital gains was increased from 25 percent to 48 percent (over a period of three years). A series of other changes followed. Many of the semiconductor executives CRA interviewed asserted that changes in tax policies in the last twelve years have made capital more difficult to obtain and restricted the ability of firms to expand capacity, enter new product areas, and grow at historical rates (interview with industry executives).

Several tax policies (Investment Tax Credits, capital-gains tax, taxation of stock options, and other tax credits) are discussed in the following sections to point out the various effects they have had on capital formation, entry of new firms (for instance, venture-capital availability), and incentives to invest in R&D. Each of these tax policies is general in nature—that is, it affects firms in all industries. None was tailored specifically to the semiconductor industry, although all of them have had important effects in this industry as well as in other high-technology industries. These policies are discussed utilizing our framework for policy analysis and highlighting the major policy changes currently being recommended by industry officials.

Investment Tax Credit

The Investment Tax Credit is available for investments in machinery and equipment, but not for buildings or R&D, and is currently set at a rate of 10 percent (AEI 1978, p. 11). Uncertainty about the Investment Tax Credit has existed since it's introduction in the early 1960s and is used in varying degrees as a countercyclical device. Until 1978, when it was made permanent, the credit was a temporary measure that was adjusted (or abolished) as policy makers believed economic conditions warranted. Studies have indicated that it may be a useful and effective incentive for business investment and that it promotes faster, long-term growth in the capital stock (American Enterprise Institute 1978).

The Investment Tax Credit may affect the firm's asset structure by providing an incentive to invest in assets that qualify for the credit (such as equipment and machinery). This incentive in turn may influence a firm's strategies; expansion of capacity may allow a firm to move into new-product areas (for example, MOS), since the credit can be applied to the equipment needed to expand capacity.

Innovative behavior may be indirectly affected since the Investment Tax Credit may leave firms with more funds to invest in R&D. Also, current proposals are under discussion (such as those by the Semiconductor Industry Association and Senate Bill S.700) to make the Investment Tax Credit applicable to R&D expenditures. This application could directly affect innovation since firms would have a greater incentive to invest in R&D (Nesheim 1979, p. 16). The stated purpose of the Investment Tax Credit is to improve economic performance by increasing productivity, creating jobs, and providing capital for expansion. Technological innovation may also occur indirectly as a result of the increases in R&D investments. Application of the Investment Tax Credit to R&D expenditures is currently an important issue with semiconductor executives and is viewed as one way of staying even with foreign government incentives to their high-technology firms (Nesheim 1979, p. 17).

Capital-Gains Tax

The capital-gains tax created a great deal of controversy among semiconductor executives during the 1970s. Between 1969 and 1976 Congress increased the effective tax rate on capital gains. The net effect was to increase the maximum rate from 25 percent in 1969 to over 49 percent in 1978. The Tax Reform Act of 1978 then reduced this maximum rate to 28 percent.

Prior to the Tax Reform Act of 1978, the capital-gains taxes were composed of three components: a standard capital-gains tax, a minimum tax, and a maximum tax benefit on personal-service income. The standard capital-gains-tax rate [Section 1202 of the Internal Revenue Code (IRC)] was increased by the Tax Reform Act of 1969 from 25 percent in 1969 to 29.5 percent, 32.5 percent, and 35 percent in the three following years. For tax payers in the highest tax brackets, the minimum tax contributed close to five points to the overall tax rate on a capital gains.

The third component, reduction in maximum tax benefits on personal-service income, contributed as much as ten points to the taxes on capital gains. Since 1971, this reduction has applied to the individuals who are in the highest tax bracket and who have some form of labor income (Section 1348 of IRC).

The Tax Reform Act of 1976 also lengthened the minimum holding-period requirement for qualifying capital gains. This period went from six months in 1976, to nine months in 1977, to twelve months in 1978 (Section 1222 of IRC).

Many economists, business executives and members of Congress expressed concern than these changes would stifle capital formation, slow growth and productivity, and prevent new, small firms from raising investment (venture) capital. This latter issue, the ability of firms to raise capital, had the greatest impact on the semiconductor industry. Because of the increased capital-gains-tax rate in 1969, individual investors in new companies faced a reduced incentive to invest in a new venture. Venture capitalists view the investment decision in terms of a risk-return tradeoff and are primarily interested in high-risk/high-return situations. A capital gain is defined as profits earned on the sale of capital assets held for twelve months or more. Thus, when a venture capitalist invests in a new firm, he expects a substantial return (usually 25 to 35 percent a year) on that investment that will be taxed as a capital gain once he liquidates the investment. However, given the increase in the tax rate, venture capitalists saw the prospective rewards of *any* investment diminish and thus were less likely to commit funds to any project, especially risky ones.

Venture capital has played an important role in the semiconductor industry, and we need to examine how policies (especially the change in capital-gains tax) have affected its availability. Venture-capital availability in every industry will fluctuate over time depending on present economic conditions and business cycles. Nevertheless, venture capital will be more readily available to firms in a rapidly growing industry such as semiconductors, as opposed to more-mature, slower-growing industries. In the former industry, the risk-return tradeoff often appears more attractive, especially during boom economic conditions (CRA 1976, p. S-7). In a strong economic environment, the venture-capital market is usually very active; however, during downturns, capital "dries up" and obtaining money for a new venture becomes more difficult.

This trend only manifested itself in the semiconductor industry during the last decade. Prior to that time, venture capital had been easily obtainable for two reasons. First, the capital costs of entry were relatively low since only a modest investment (about $1 million) was required to set up an efficient manufacturing operation. Thus, there was less demand for large sums of venture capital. Second, the rapid pace of technological change and growth in the industry during the 1950s and 1960s made the profit opportunities for venture capitalists good and the risks relatively low (interview with venture capitalists).

However, in the late 1960s and early 1970s venture capital became more difficult to raise for new semiconductor ventures. This difficulty was due in part to the change in the capital-gains tax (interview with venture capitalists). Also, the capital costs of entry had risen considerably, thereby requiring larger initial outlays. Estimates of capital costs of entry range from $100,000 to $1 million in the 1950s to $10 million in 1970 to $30-40 million today for a state-of-the-art manufacturing facility (*Forbes*, 26 November 1979, p. 56; Tilton 1971, p. 88). The costs to set up a wafer-fabrication plant currently run about $10 million and may rise as high as $30 million by the mid-1980s (*Business Week*, 3 December 1979, p. 81). These increases (and the resulting decrease in entry) stem from the increasingly complex and costly technology required to make each new device. As a result, venture capital has not flowed into the industry in part because entry costs are too high to make a venture-capital offering feasible. Finally, the return on venture-capital investments dropped and several financings turned sour (interview with venture capitalists).

Semiconductor firms often have difficulty raising growth capital. Several firms of moderate size have elected to merge with larger firms to help solve their capital problems. Firms may be forced to alter their strategies when faced with a capital shortage. For example, Signetics entered the industry in the 1960s with venture capital from an investment banking house and followed a strategy of moderate to rapid growth. However, within three years additional capital was needed and Signetics decided to merge with Corning Glass, which was willing to provide the capital needed for expansion. Later in the early 1970s, the Dutch electronics firm, Philips, acquired Signetics and infused further capital into the company. Another foreign firm, Siemens A.G., acquired a 20-percent interest in AMD and also set up a joint venture with AMD in the United States (*Electronics*, 13 October 1977).

As seen in previous chapters, many of the major integrated-circuit innovations came from firms spinning-off from existing firms (for example, Fairchild, Signetics, Intel, Mostek). If firms are unable to enter the industry due to a lack of capital, innovative pressures will be reduced on the firms already in the industry.

Recent reports indicate that the venture-capital market may be booming once again. Recently, evidence indicates that venture capital is available, particularly for rapidly growing firms. Little money is going to firms for start-ups that require the development of a new technology (*Electronics*, 10 May 1979, p. 88). But venture capital is flowing to new firms adapting or improving advanced technologies where profit potential appears high (interview with venture capitalists).

One explanation for this increased availability of venture capital was the change in the capital-gains tax in 1978. The net effect of the 1978 change was to reduce the maximum capital-gain-tax rate from 49 percent to 28 percent. The minimum tax and the reduction in maximum tax benefits would no longer apply to capital gains; these taxes would be replaced by an alternative minimum tax that would not increase the maximum tax rate over 28 percent (Section 55 of IRC). Also, the maximum amount of profit subject to the tax was dropped from 50 percent to 40 percent.

Several groups (Walker and Bloomfield 1979) pushed hard for this reform and in the year that has passed since the tax reduction, many people believe it has stimulated capital investment (*Wall Street Journal*, 31 October 1979) and in particulr venture-capital offerings Mick 1979, p. 28; interview with venture capitalists). Data indicating increased activity in the venture-capital market include an increase in the number of equity offerings by firms with net worth under $5 million from 15 in 1978 to 34 in 1979 and an increase in the funds raised by such firms from $76 million in 1978 to $166 million in 1979 (*Wall Street Journal*, 10 December 1979). Other people believe that the so-called demonstration effect of earlier successful investments has helped stimulate the flow of venture capital. Venture capitalists are now more experienced and more willing to pass up marginal deals (interview with venture capitalists). At any rate this decrease in the capital-gains tax may stimulate new investments and provide more capital, which may ultimately result in more R&D spending and improved innovative performance.

Treatment of Stock Options

The tax treatment of stock options and the elimination of the qualified stock option in 1976 have influenced the ability of semiconductor firms to attract top personnel from other firms. Since skilled-labor resources have historically been in short supply in this industry, firms have used the stock option as a means of luring top-notch personnel to new firms. Employee-ownership plans, the profits of which were taxed at a relatively low rate, were viewed as an attractive incentive to engineers and were beneficial to a new firm in obtaining qualified personnel. These plans made

Analysis of Government Policies

possible large incomes contingent upon the success of the new firm and allowed taxation at the capital-gains rate, which was lower than the personal income-tax rate.

Changes in the tax laws now make these options less attractive to employees and new firms. Previously, qualified options were given to key personnel, and no tax was assessed either at the time of the grant or at the time of exercise. At the time of resale, the shares were taxed at the capital-gains rate. These special options were phased out beginning in 1976, and stock options no longer receive special tax-exempt treatment (U.S. Senate 1978, p. 43).

Firms now prefer to use other employee-incentive plans such as bonuses and profit sharing; however, none is as potentially lucrative as was the qualified stock option. For example, one firm, Signetics, has a profit-sharing plan in which employees receive a bonus based on a progressive rate structure and the relationship of the company's actual profits to a predetermined benchmark for the year. Thus, if the base is 30 percent for the year and Signetics earns 40 percent that year, the amount that Signetics contributes to the profit-sharing pool will increase more than proportionately to the increase in profits above the base. When the firm's pofits fall short of the base, its contribution to the profit-sharing pool also drops more than proportionately, according to a CRA interview with an industry executive. (The figures used in this example are entirely hypothetical and are not related to the figures in the Signetics profit-sharing plan.)

Semiconductor executives believe a return to the qualified stock option would benefit the industry (interview with industry executive). These options had an important impact in attracting people to new firms, making it easier to start a new firm. This effect on entry is particularly significant given that new, small firms have been an important source of innovation in the industry (see chapters 3 and 4).

Other Tax Policies

Recent changes in the U.S. income-tax policy have eliminated the favorable treatment for Americans working in foreign countries for U.S. firms. Overseas personnel no longer earn their first $25,000 tax free. As a result companies find that convincing top-quality management to live outside the country is difficult. Concern has been expressed over the effect of this problem on the growth of the semiconductor industry. One official stated that the new policy was "excessively punitive," and that the number of U.S. semiconductor-business personnel living abroad had dramatically decreased as a result of the policy change. He stated that companies are finding it more difficult to penetrate export markets because of the scarcity of the requisite-management expertise within the native labor force (interview with industry executive).

Another proposed revision of the tax laws might also affect adversely productivity and R&D spending. The proposed change in Section 1.861-8 would limit R&D deductions on the firm's tax return to the same proportion on earned income as to worldwide income. This limitation could deny firms important R&D tax advantages and or force them to move some of their R&D abroad. This move could conceivably make firms' R&D efforts less effective since the exchange of research information may be hampered. More importantly, industry officials believe that the incentive this law would create to "export" the development of new technology runs counter to national priorities and needs.

Manpower Policies

The semiconductor industry is best characterized by its people rather than its companies, since company success directly depends upon individual expertise (Braun and MacDonald 1978, p. 128). Therefore, government policies that affect the quality and availability of manpower in the industry have an important impact on industry growth and structure. We discuss now the role the government has played in training engineers for the industry and encouraging or discouraging their mobility within the industry. Issues include government funding to universities; U.S. income-tax policies with respect to the treatment of stock options, overseas operations, and tax credits for R&D funding; antitrust regulations; and several court rulings regarding interfirm mobility.

Government funding, particularly from the DOD and NASA programs, greatly aided the electronics industry in the 1950s. During that period the DOD allocated $1 to $2 million annually to more than 100 doctoral candidates for basic research in solid-state electronics (Utterback and Murray 1977, p. 17). Universities received funding for the necessary facilities and equipment to attract top-quality research personnel for government projects. The promotion of university research and training provided the highly technical labor force needed for the development of the commercial industry. Expertise gained on government projects could be applied to the growing demand for consumer applications. This was particularly important in facilitating the growth of small firms and encouraging new entrants.

Since the 1950s and mid-1960s, government support for R&D and manpower training to universities and firms has markedly declined (interview with industry executive). Several major firms claim that support is currently nonexistent. Many executives have agreed that the universities provide an important training ground for the technical manpower required by the industry. Generally, however, executives believe that university facilities and equipment are obsolete and hinder up-to-date training of future engineers (interview with industry executive).

Several semiconductor executives believe the government should increase its funding to universities to help update their engineering facilities and programs. They believe that enrollment in engineering programs is directly correlated with the fluctuations in government funding (Utterback and Murray 1977, p. 37). Nevertheless, they also believe that the quality of personnel attracted to the industry has not diminished over time (interview with industry executive).

A new tax issue is pending before Congress that many executives believe could have a positive effect on the availability of skilled manpower in high-technology industries, including the semiconductor industry. Senate Bill S.1065 would provide for a federal tax credit of 25 percent for grants by U.S. corporations to colleges and universities for fundamental research. The industry contends that the proposal would positively affect technical innovation by encouraging the expansion of the base of fundamental knowledge that supplies the building blocks for advancement in the industries of applied science (Nesheim 1979, p. 19). In addition, the new policy would increase the supply of talented, highly skilled engineering students who are the future innovative leaders in U.S. business.

Litigatory proceedings have limited interfirm mobility to some degree. Despite the difficulty in obtaining an injunction against departing employees, actions were sought in a few cases. In 1969 IBM went to court against Cogar Corporation, which acquired 70 of its 150 employees from IBM. Cogar was enjoined from further recruiting IBM personnel. Motorola took Lester Hogan and other former company executives to court when they joined Fairchild in 1968. Fairchild brought action against the founders of Rheem Semiconductors, alleging that they had stolen proprietary material. National obtained a consent decree Jean Hoerni and Intersil, After Hoerni recruited a group of experts from National when he formed Intersil.

These court cases define the limit of tolerance in the industry for so-called employee theft. Except in these extreme cases, when at least several top employees are lost to another firm, semiconductor companies have had no legal means of slowing down the high rate of manpower mobility. Few top people have worked for only one firm and the vast majority have worked for three or four firms within a decade (Braun and MacDonald 1978, p. 131).

In sum, the supply and mobility of manpower were initially promoted by government policies, especially with university funding for research and training and antitrust action. Government actions have faltered, however, and several major firms believe that the cut in university funding and new tax policies concerning stock options and overseas income have limited the availability of highly skilled technical manpower.

Policy Interdependencies

As we discussed in chapter 1, policies cannot necessarily be examined independently. Consideration must also be given to their interaction and interdependence. Interdependencies should be distinguished from simultaneous impacts. Interdependence implies interaction among policies whereas simultaneity implies concurrent but independent impacts of policies. These policy interactions may alter the anticipated effects of one or a group of policies. Also policies may act together to reinforce one another. These interactions and interdependencies as they relate to the semiconductor industry can best be explained by examining them historically. In this way further light may be shed on the changing role of policies over time and the reasons certain policies have affected the industry so strongly at certain times. Not all interdependencies and policy interactions can be discussed here; however, the most important ones will be described and analyzed to illustrate how our framework can be applied to them.

Interdependencies arise in several ways. A policy may affect more than one element of the framework; these effects may be contradictory or reinforcing. If the impacts are contradictory, then the results may not be the anticipated ones. Also, multiple policies may each affect only one element of the schema, but taken together these effects may be other than what was anticipated. Finally, a policy may be implemented with the expectation that a certain element of the framework will be affected; however, for a number of reasons the impact may not be realized.

During the brief history of the semiconductor industry, several instances have occurred in which policies have interacted and had a significant impact on the industry. The most obvious example was government R&D-funding and procurement activity that during the 1950s and early 1960s interacted to create a conducive environment for innovation. Other policies have been interactive: antitrust and patent policies have had a synergistic effect in making possible easy diffusion of technical information, and certain tax policies have had important interactive impacts.

R&D funding and procurement worked at cross purposes concerning the entry of new firms but reinforced one another to accelerate the rate of innovation in the industry. During the 1950s, government R&D funds went primarily to the established firms and only infrequently to newcomers. One explanation for this distribution may be that contracting officers awarded

R&D funds based on past performance, since the achievable research was done. Newer firms with less-recognized research staffs could not compete as effectively as established firms with a known track record.

In contrast to R&D policy, early government-procurement policies favored newer firms, and established firms did not receive a proportional total of procurement funds (Tilton 1971, p. 91). This differential impact may be explained by the willingness of new firms to adapt more readily to government specifications, since they were trying to establish a competitive base. Also, past performance would be less important, since the output of a procurement contract can be well specified, compliance with contractual requirements can be easily monitored, and new firms may have the innovation to fit the government's needs.

Although these two policies worked at cross purposes concerning entry, they acted together to create a conducive environment for innovation. Procurement policy tended to reduce the market risk faced by the potential innovator and R&D funding reduced technical risk. These policies together had an impact on almost every firm in the industry and as a result, the firms in the industry were responsive to government demands for new, better products. These policies also interacted to create a so-called urgency impact for many of the firms, compelling them to match and increase their R&D and innovative efforts (Kleiman 1966, p. 68). Both policies impacted strongly on the environment of firms and forced them to match their strategies to the government's demands. The magnitude of the government's efforts in this area were such that firms could not ignore these incentives.

Antitrust and patent policies have had interdependent effects. The threat of strict antitrust enforcement, especially during the 1950s, prevented firms from using patents as a means of acquiring or maintaining market share. Royalties tended to be relatively low and cross licenses were readily obtained; otherwise firms (especially the large receiving-tube firms) believed they might have risked a suit with the Justice Department (interview with industry executive).

In this case both policies acted simultaneously *and* interdependently on the environment and in turn affected the way firms formulated their strategies. The larger firms in the industry had to be more careful about obtaining a dominant patent position or trying to enforce those patents. The two policies interacted since they both influenced the firm's environment and *together* forced firms to reconsider and possibly set up new strategies. Had the threat of antitrust enforcement not existed or if the opportunity for obtaining a strong patent position had not presented itself, firms would not have encountered such obstacles and might well have pursued very different strategies with respect to licensing and enforcement of patent rights.

In this case the policy interaction mitigated the anticompetitive effects of patents while preserving the benefits of patents by providing and diffusing new information. However, this impact was beneficial to the industry only as long as all firms felt constrained by antitrust policy. Today, foreign firms seem to be catching up and have built up strong patent positions. The constraint imposed by antitrust policy now has a differential impact, affecting only domestic firms. In the long run foreign firms may be able to use their patent position to gain a large market share in the United States (interview with industry executives).

One example of a simultaneous policy impact was the tax treatment of stock options and capital gains during the 1960s. Entry of new firms was facilitated by liberal rules regarding stock options. At the same time, capital-gains taxes were relatively low and venture capital (especially in the late 1960s) was readily available and facilitated entry (interview with venture capitalists). These tax policies together allowed easy entry of new firms, which increased the innovative and competitive pressures in the industry. Our framework shows this impact quite readily. Two of the crucial inputs into forming a new semiconductor firm were (and still are) highly skilled engineers to develop innovative products and investment capital to set up the business. The qualified stock option and a moderate capital-gains tax operated to aid entrepreneurs interested in forming a new firm to obtain both of these inputs. Qualified stock options made it financially easy to attract skilled labor by offering an attractive stock-ownership plan; moderate capital-gains taxes helped keep venture capital flowing, thus aiding in setting up a new firm.

Analysis of the VHSIC Program

This section presents the important questions policy makers should ask when analyzing the impact of the government's VHSIC program on the behavior and performance of the semiconductor industry. We will not deal with *all* of the questions outlined in table 1-1; only those questions that would reveal the most information about this policy will be presented, together with reasons why these questions are so important. Answers to the questions will *not* be given since the purpose of this section is merely to indicate how the framework allows policymakers to ask relevant questions.

The VHSIC program, as we mentioned, is an attempt by the DOD to fund a major research effort in the field of electronics and integrated circuits. The research effort is anticipated to last six years and total spending is expected to reach $200 million.

With regard to the environment, several important questions must be

Analysis of Government Policies 175

asked about the VHSIC program. On the demand side, one question is whether VHSIC will lead firms to alter their output mix. This aspect is important given that the output for military use is specialized and differs somewhat from commercial devices. The costs of changing one's output mix may have already led to opposition to the policy. On the supply side several questions are relevant. For example, how will the policy affect the technology base of the industry? Will it generate new technical discoveries? Also, how does the policy affect the opportunities for innovation and the technical risk faced by firms in the industry? What is the opportunity cost in terms of commercial innovation foregone? These questions are important since they relate to whether the policy will generate and allow the creation and adoption of new technologies.

Two other questions are whether the policy will affect the supply of skilled resources to the industry and whether VHSIC will affect the international competitiveness of U.S. firms. The first question is important, especially given the shortage of certain skilled resources in the industry. VHSIC may be one way to increase the supply over the long run, but in the short run the shortage may become more acute given a diversion of resources to the VHSIC program. VHSIC's contribution to international competitiveness is also important, since one of the implied goals of the program is to keep U.S. firms ahead in the technology race. VHSIC is also viewed by many industry members as a response to the massive funding efforts by foreign governments; thus, the impact of VHSIC on international competition is an important consideration.

One final question about the impact of VHSIC on the environment is whether participating firms will be able to appropriate the returns from innovations generated from the program. This question is important because if firms do not believe they can realize sufficient returns, then they will be unwilling to bid on the program. Also, firms must believe that potential innovations or other benefits are to be gained from engaging in this program.

Moving to questions relating to other dimensions of the framework, questions about VHSIC's impact on firms' corporate goals and organization do not appear to be significant, with the exception of the impact of VHSIC on a firm's ability to alter the mix of skills in its resource base. This impact is significant for firms wishing to bid on the VHSIC program but with little prior experience in research and production of military devices. If the policy allows firms to alter their resource base by adding new engineers or investing more in R&D to accomodate these changes, then more firms will have an incentive to bid on the project.

This links us to the questions dealing with strategy formulation. These tend to be the most important questions. For example, will VHSIC reinforce or alter existing mixes of strategies? Will firms following an incremental-innovator strategy view VHSIC as a means of becoming a key innovator? How does VHSIC alter the incentives for innovation at the firm level? These questions are vital for understanding the impact of VHSIC since the answers will demonstrate how firms react and respond to the policy given its initial impacts on the environment. Some firms may try to pursue new strategies and achieve success as a result. Other firms may fail either by continuing with established strategies or by following a new set of strategies.

The other set of important questions involves the reactions of rival firms to changes in the mix of strategies of one of more groups of firms. And how does this reaction affect the number, size, and distribution of strategic groups in the industry? The answers to these questions will tell the policy maker a great deal about the competitiveness of the industry, the likely effect on innovative behavior, and strategy formulation. And perhaps more importantly from the policy maker's perspective, these answers lay the foundation from which the effects on performance can be posited.

Through its impact on the environment, strategy formulation, and competitive behavior, VHSIC may have an impact on performance. The questions to be asked are whether VHSIC will affect social performance—that is, will it change the technology, prices, costs, quality, and diversity of product offerings in the industry? Again, one of the assumed goals of the policy is to improve social performance by improving the technology and hence the products offered. These questions, while somewhat speculative prior to completion of the program, still provide important insights into the potential for improving social performance.

Other important questions are related to the spillover generated by the VHSIC policy and how firm performance may be improved by the policy. Firm performance, such as profits or growth, may be affected, especially firms' expectations of these elements. If the firm believes expected profits or growth will be enhanced by the policy, then it may be more willing to undertake the policy. Spillovers, or the potential for spillovers, are also important for social performance. A policy that generates beneficial spillovers will be more socially beneficial and therefore more desirable.

These are some of the questions policy makers should ask when analyzing a prospective policy such as VHSIC. By proceeding through the questions and determining which are most important, the answers generated will provide timely information regarding the impact on firm behavior and economic performance. The framework provides an easy and comprehensive tool for understanding and analyzing how a policy will affect an industry.

Analysis of Government Policies

Summary and Conclusions

This chapter illustrates how the framework for policy analysis discussed in chapter 1 and in CRA (1980) may be used to assess and analyze the impact of a range of policies on a firm or industry. The important impacts of various policies affecting innovative behavior in the semiconductor industry have been analyzed. We have found that policies can have differential effects among firms in an industry and that the importance and effects of policies change over time as the industry evolves. Finally, the framework for policy analysis has been used to illustrate the independent and interdependent impacts of policies on firm behavior (particularly innovative behavior) and performance in the industry, and how policies affect various strategic groups in different ways. We have sought to provide the policy maker with examples of how to analyze the potential impacts of a policy on market structure, innovative behavior, competitive strategies, and firm and industry performance. We have used our framework for policy analysis to trace the linkages of a policy from element to element and ultimately to its effects on performance. The framework enables one to discern how a policy such as procurement stimulated innovation: procurement reduced the market risk to firms of introducing a new product, which induced them to introduce a new device at a premium price, realize learning economies, move into new applications, improve performance, and have the resources to innovate again.

Several conclusions can be derived from our analysis of government policies affecting the semiconductor industry. Policies impact various firms differently and these impacts will change over time as the industry itself changes. Policy makers must bear these points in mind when designing and implementing policies. Government R&D funding was important during the early years of the semiconductor industry because it lowered the technical risk firms faced in developing a new technology. However, not everyone has welcomed this type of funding (nor did they in the 1960s when Fairchild refused government R&D funds); it may not lead to the "right" technology in the future and may divert resources away from more-promising endeavors. The impact of the policy may be different in the future if some of the leading innovators in the industry do not participate in the government programs. Instead of stimulating further R&D and innovative activity, it may cause firms to spend less on promising commercially oriented research programs and instead follow the government's research priorities.

We have also seen the firms react differently to policies based on their strategic mix and resource base. Firms following major-innovation strategies may not believe in government intervention in the face of foreign competition, whereas firms pursuing incremental-innovation strategies are more threatened by foreign competition and favor government intervention.

Policies have affected firms and their innovative behavior both directly and indirectly. Often it may be unclear whether the direct or the indirect impacts are stronger or more important in determining a firm's innovative behavior. A direct impact occurs, for example, with government R&D funding, which provides capital for research and may stimulate firms to invest more in R&D. An indirect impact occurs with tax policies such as stock options, which affect interfirm mobility and the ability of new firms to attract top-notch personnel. Without this input, many new firms might not have succeeded and innovation in these firms might not have occurred.

Most of the policies that have affected the semiconductor industry have been general policies with indirect effects on innovative behavior and performance. Government policy during the 1950s and 1960s created a conducive environment that in turn allowed individuals to be creative and develop new products. During the 1970s other policies have indirectly affected innovation by reducing labor mobility and slowing capital formation. Key personnel and capital are two of the essential ingredients for successful innovation, and policies that have affected them may have indirectly retarded innovation in U.S. firms.

Thus, in order to understand policy issues affecting the semiconductor industry today, the fact that the indirect impacts of policies can be as important as the direct impacts is significant. Trade policy, which is of current concern to many industry executives, does not appear to affect innovative behavior directly; however, if sheltered foreign markets give foreign firms increased cash flow, they may outdo U.S. firms in the area of incremental innovation. Tax policies are also of current interest; tax policies designed to make capital more available and to better reward the successful innovator can often have important indirect impacts on innovative behavior. Policy makers should be aware of these indirect effects as well as the direct effects of a government program such as VHSIC, since these indirect impacts may often be as important to innovative behavior as the direct effects.

Appendix A
Glossary of Integrated-Circuit Technology

This glossary contains brief descriptions of the innovations listed in tables 3-1 through 3-3. For further explanation of integrated-circuit technology, see the second section of chapter 2.

Major Semiconductor Device Advances

Integrated Circuit This circuit was a fundamental conceptual advance involving the placement of more than one component in a single semiconductor body. The technique has evolved from a few devices per chip in the early 1960s to 35,000 components or more in today's most complex products.

Diffused Resistor The diffused resistor was the first component, other than diodes or transistors, to be incorporated in integrated circuits. Useful digital circuits were possible using only transistors, diodes, and diffused resistors. Diffused resistors are formed using the same processes that are required for the formation of diodes and transistors. This compatibility was important in making early integrated-circuits practical.

Metal-Oxide-Semiconductor (MOS) Capacitor The gate structure of a MOS transistor is itself a capacitor. This characteristic is very useful in MOS-circuit design. Most modern memories use the capacitance of the MOS structure as the basic storage mechanism.

Metal-Oxide-Semiconductor (MOS) Transistor This was the basic MOS structure consisting of a semiconductor body (silicon) with silicon-dioxide gate dielectric and metal (aluminum) gate. This structure, with variations, remains the basic MOS transistor to date.

Substrate-Diffused-Collector Transistor This structure is the basic bipolar integrated-circuit transistor. It was not the original integrated-circuit-transistor structure, a triple-diffusion transistor, but has entirely replaced the early structure. The advantages of this structure are less high-temperature processing, improved electrical parameters, and less surface area.

Complementary Metal-Oxide-Semiconductor (CMOS) Transistor The basic MOS transistor can be either n-channel or p-channel. CMOS combined the two types in a circuit form that has advantages despite more-complex processing and relatively large surface areas. The advantages are low power consumption, high noise immunity, and high speed.

MOS Resistor This resistor is actually an MOS transistor. Properly designed and biased, it serves satisfactorily as a resistor in MOS circuits. Its advantage is that it can be formed at the same time that the transistors are formed, thus minimizing processing.

Deposited-Metal Resistor This type of resistor gives greater precision and better temperature coefficients than do diffused resistors. It can be placed on a nonsemiconductor body, as in hybrid circuits, or on top of the semiconductor chip.

Depletion-Mode MOS Resistor Most MOS transistors operate in the enhancement mode—that is, the transistor is "off" with no gate bias. A depletion-mode transistor is "on" with no gate bias and is turned off by gate bias. As a resistor it has superior performance as compared with enhancement-mode MOS resistors.

Bipolar-Junction Field-Effect (JFET) Combination This technique combines conventional transistors (emitter-injection) with field-effect transistors to provide circuits with the best features of both. The problem was to develop processes compatible with both devices.

Trapped-Charge Storage A problem of semiconductor memories is volatility—that is, the memory contents are lost if power is removed. This problem can be alleviated by setting the memory cells in relatively fixed states through the use of trapped-charge layers. Such structures can retain the contents of the memory for significant periods of time after power is removed.

Bipolar-MOS Combination As with JFET combinations, this technique combines two transistor structures in a single semiconductor body in order to get the best features of both.

Isoplanar, Coplamos et al. A problem with early MOS devices was the relative nonflatness of the surfaces that resulted from differing thicknesses of silicon dioxide on the surface of the chip. These variations were by-products of the processing cycle and can cause problems with the metallization patterns. The subject techniques give a relatively flat surface with improved yield and reliability.

Merged Transistor (I^2L) This structure is area efficient by not having each transistor as a separate component. A group of transistor collectors are fed from a common emitter.

Vertical MOS (VMOS) Reduction of MOS-transistor surface area and enhanced operating frequency is possible by using diffusion techniques to form the source, drain, and channel. This is accomplished by etching a V in the surface of the semiconductor and placing the gate structure on this sloped surface. These benefits are gained at the cost of increased processing complexity.

Major Semiconductor Process Advances

Resistive Metal Deposition This process involves the deposition of metallic films that are characterized by having significant electrical resistance, so that the films can be used as resistors in electrical circuits. This process is in contrast to metallic films that are also deposited but serve only as electrical connections between components. Deposited resistors are superior to diffused resistors in that their tolerance can be controlled more tightly.

Ultrasonic Bonding Ultrasonic bonding is a technique that uses sonic energy to attach an aluminum wire to an aluminum substrate or "pad" on a semiconductor chip. The ultrasonic movement breaks the surface oxide of the aluminum and allows a true bonding of the underlying aluminum. Its advantage over gold-aluminum systems is in the avoidance of the formation of intermetallic gold-aluminum compounds that can, dependent upon time and temperature, weaken or break the bond attaching the wire to the substrate or chip.

Annealing for Stress Relief Annealing is an old metallurgical technique used in semiconductor processing to relieve crystalline stresses resulting from high-temperature processing. Unrelieved crystalline stresses degrade the performance of the resulting semiconductor devices.

Gettering Processes Gettering processes involve the use of one element in a semiconductor body to immobilize another element. This process is typically used where impurities would degrade device performance if allowed to remain mobile. Gettering prevents this.

Plastic Encapsulation Plastic is used in semiconductor-device packaging purely as an inexpensive packaging. It replaces a metal can or ceramic package and is usually molded.

Direct Bonding (Bump) Originally developed by IBM to allow automated assembly of alloy semiconductor devices, this technique has been modified for "beam lead," "gang bonding," and other approaches that have in common the simultaneous attachment of multiple connections to the semiconductor chip.

Nitride-Chemical Vapor Deposition The use of deposited nitride layers was originally conceived as a dielectric in the gate structure of MOS devices. It is now used more frequently for masking purposes in forming MOS-device structures.

Glass-Chemical Vapor Deposition Amorphous glasses are deposited on completed semiconductor devices as a protective layer over the soft aluminum interconnection. When properly doped with phosphorus or other elements, it can also aid in the gettering of impurities.

Ion Implantation This process accelerates ions to high energy levels and then directs them to the surface of semiconductor wafers. The ions penetrate the semiconductor body to varying depths depending upon the acquired energy. When left in the surface oxide layer these ions serve to adjust oxide and surface charge concentrations. When penetrated into the semiconductor body the ions serve to modify the type of conductivity in the formation of device structures.

Schottky Junction The Schottky Junction forms a diode composed of an underlying semiconductor body and two or more layers of deposited metal. It is normally used in conjunction with a transistor and serves to keep the transistor from entering saturation, thereby improving the performance of the circuit.

Dielectric Isolation Electrical isolation of components in bipolar integrated-circuits is normally accomplished by means of intervening reverse-biased diffused junctions. This isolation technique adds unwanted capacitance to the circuit and decreases its performance. Various dielectric-isolation techniques have been developed that isolate components by separating them with low-conductivity dielectric materials.

Sputtering Sputtering is an old process in which high fields and ions are used to energize one material so that it is transferred, in vacuum, to form a coating on a target. It is an alternative to evaporation or ion bombardment. It has the advantage that two or more materials can be sputtered at the same time, thereby forming reacted or composite coating.

Polysilicon Deposition This process deposits layers of noncrystalline silicon. These layers, when delineated, serve as conductive paths and/or as part of the gate structure in MOS devices.

Silicon-on-Sapphire This process uses a sapphire substrate upon which a silicon film is epitaxially deposited and subsequently preferentially etched into individual devices. Its advantage is improved performance as compared to conventional MOS devices.

Plasma Etching Semiconductor processing has historically used liquid chemicals for various etching and material-removal procedures. Plasma etching accomplishes the same result through the use of ionized gases, primarily fluorine compounds.

Ion Milling This process uses streams of ions to remove materials from a target surface. It is an alternative to chemical etching and other removal techniques.

Electron-Beam Processing Electron beams can be used for a variety of purposes in semiconductor processing. Of most use to date are beams to expose photosensitive lacquers in the manufacture of masks for use in very fine-line photolithography. In the future electron beams will be used in advanced photolithography systems to expose directly the photosensitive resists in semiconductor processing.

Photolithography-Projection Alignment Conventional mask-alignment systems have had the drawback of either destructive contact between the semiconductor material and the mask or diffraction limitations when the mask is moved out of contact. This technique uses the scanning of mirrors to allow good resolution with significant separation between the mask and the semiconductor body.

Deep Ultraviolet Photolithography Better photolithography resolution can be realized through the use of shorter-wavelength light during the exposure process. Deep ultraviolet light has a shorter wavelength than do the conventional ultraviolet sources. This light may allow some advantage in small-geometry circuits if a suitable lens can be developed.

Major Product Families

Logic Families Logic devices are used in digital applications to perform arithmetic functions and make decisions. The names of the logic families describe the essential components or means of connecting the components to form the basic logic gate or switch. Many gates are connected (in a variety of circuit designs) to perform logic functions such as adding or shifting data. The families shown in table 3-3 are:

RTL = Resistor-Transistor Logic

DTL = Diode-Transistor Logic

ECL = Emitter-Coupled Logic

TTL = Transistor-Transistor Logic

DTL provided superior noise immunity compared to RTL. ECL is a very high-speed family that has been used mainly in the most powerful and advanced computers. TTL provided high speed compared to DTL but not as high as ECL. A later development, Schottky TTL, provided increased speed (but still not as fast as ECL) without as great an increase in power consumption.

Memory Families Memories store binary data and are classified by their access properties and by storage capacity. One kilobit (1K) of storage capacity equals 1,024 bits of binary data.

RAMs are random-access memories and provide temporary storage capacity. Data can be put into or retrieved from any particular location in the memory as easily as any other location (thus the term *random access*, as opposed to *serial access* in which locations can be reached only in a specified order).

ROMs are read-only memories that provide permanent storage such as for tables of data or characters for display.

PROMs are programmable read-only memories in which the user can decide what data are to be permanently stored.

EPROMs are erasable programmable read-only memories that allow the user to erase stored information and replace it with new information.

Microprocessors Microprocessors provide all the central processing-unit functions of a computer in a single integrated-circuit. They are referred to by the length of word in which calculations are processed (for example, 4-bit, 8-bit, and so on). A microprocessor combined with memory and input-output integrated-circuits becomes a microcomputer.

Appendix B
Financial Incentives in the Semiconductor Industry: The Pot of Gold that Few Ever Reached

> Promotion above the salary ceilings that often exist in large companies may be slower and more difficult than starting a new company or joining an existing small firm in which salary ceilings (or floors for that matter) do not exist. Some individuals in the industry earn a great deal of money and a few have made fortunes (Braun and MacDonald 1978, p. 131).

> The semiconductor world is populated by almost-rich professional managers who took jobs with a nice salary expecting to become millionaires with the stock option they got, and every one of us has somewhere in the dresser drawer stock options worth garbage and not that much money in the bank (Braun and MacDonald 1978, p. 132).

These two statements—the former lauding the financial opportunities, the latter describing a sobering personal experience—cover the full range of attitudes with regard to financial rewards in the semiconductor industry.

A careful examination of company and public documents indicates that few individuals have actually achieved unusual financial rewards—that is, rewards much greater than they could have achieved in some other pursuit or profession. For each Robert Noyce and Gordon Moore who was successful with Intel, countless others attempted and failed. Of course, maybe the combination of doing as well financially as elsewhere, having a long shot at becoming truly wealthy, and working on a rapidly changing technology provide just the right ingredients for strong individual effort. In any case, the prospects of becoming wealthy do appear to be a long shot, since so few individuals have actually made it.

This appendix focuses on the leading firms in the industry. It examines management stockholding in each firm and direct compensation of senior officers. Combined management stockholdings never exceed 6 percent of total stock value in any company except for Motorola (close to 10 percent) and Intel (close to 20 percent). In no semiconductor firm does management compensation stand clearly out of line with that of the electrical-equipment industry or the manufacturing industry in general (see table B-1, figures B-1 and B-2). In the same way, white-collar salaries in the semiconductor industry are clearly in line with salaries prevailing in other comparable in-

---

Louis Caouette wrote appendix B.

Table B-1
Total Direct Compensation of Senior Officers at Selected Semiconductor Firms: Fiscal Year 1978
(thousands of dollars)

Firm	Direct Compensation for the Highest-Paid Officers					Average for Next Group of Officers	
	First	Second	Third	Fourth	Fifth		Number
AMD	225	129	97	—	—	84	5
	(100)	(44)	(32)			(23)	
American MicroSystems	185	90	82	—	—	59	7
	(60)	(21)	(14)			(11)	
Fairchild	383	178	138	123	123	89	15
Intel	236	214	189	169	100	98	8
Intersil	215	126	124	—	—	—	—
Mostek	191	168	114	106	98	90	83
Motorola	371	290	240	204	202	(24)	
	(140)	(110)	(100)	(66)	(90)	80	11
National Semiconductor	156	123	123	—	—	119	52
Texas Instruments	573	511	335	282	253	(39)	
	(302)	(277)	(166)	(161)	(114)		

Source: Calculations by CRA 1980, from annual proxy statements of semiconductor companies, fiscal year 1978.
Note: Total direct compensation equals salaries plus incentive compensation (figures in parentheses). The definition of incentive compensation varied across companies, and its value was included only when it was a significant portion of salaries.

Appendix B

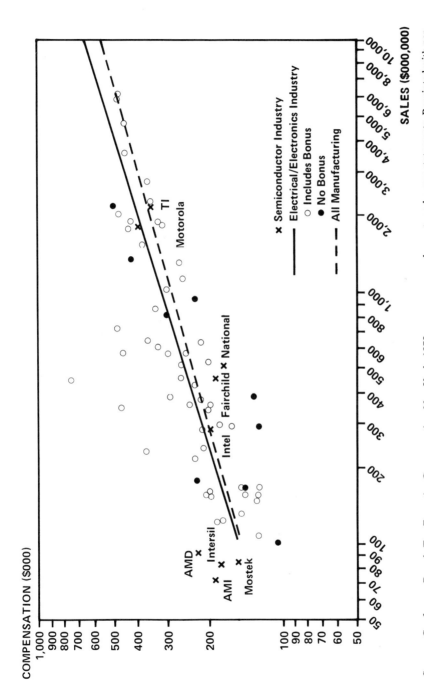

Source: Conference Board, *Top Executive Compensation*, New York, 1978: company annual reports and proxy statements. Reprinted with permission.

Note: The Conference Board trend lines start at $100 million in sales. Data are for calendar year 1977 or closest fiscal year.

Figure B-1. Compensation of the Highest-Paid Executive: Semiconductor Industry versus Electrical and Electronics Industry (SIC–36) versus All Manufacturing Industries

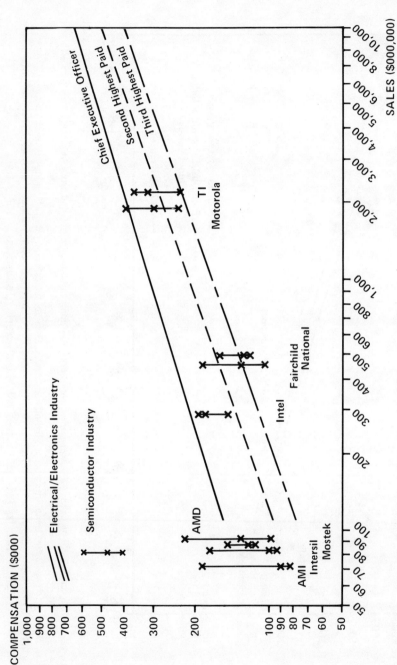

Source: Conference Board, *Top Executive Compensation*, New York, 1978: company annual reports and proxy statements. Reprinted with permission.

Note: The Conference Board trend lines start at $100 million in sales and are fitted through a scatter of data for sixty-five companies. Data are for calendar year 1977 or closest fiscal year.

Figure B-2. Compensation of the Three Highest-Paid Executives: Semiconductor Industry versus Electrical and Electronics Industry (SIC-36)

Table B-2
Average Annual White-Collar Salary in Selected Industries

Industry	Salary in Dollars[a]			Salary as a Percentage of Semiconductor Salary		
	1972	*1975*	*1976*	*1972*	*1975*	*1976*
Telephone and telegraph apparatus	12,590	15,457	17,304	92.4	91.4	91.7
Radio and television communications equipment	13,616	16,389	17,851	100.0	96.9	94.6
Semiconductor	13,620	16,916	18,870	100.0	100.0	100.0
Motor-vehicle and car bodies	16,197	19,734	22,667	118.9	116.7	120.1

Source: Calculated by CRA 1980, from 1972 Census of Manufacturers and 1976 Annual Survey of Manufacturers, Bureau of Census, Department of Commerce.

[a] Average white-collar salary equals average salary for all employees minus average salary for production workers.

dustries (see table B-2). Proxy statements which are a principal source of data for this appendix, provide information on the most highly paid individuals and major stockholders. As such, they set the upper limit of financial rewards within any given firm. Even though specific information is not available for other individuals within the firm, their total individual compensation will, except in unusual cases, be less than the compensation of officers mentioned in official documents.

Types of Incentives

Financial incentives have three major dimensions: how they are determined, what they amount to, and who gets them. These financial incentives can be tied to the financial results of the company as a whole, or can be tied to more specific events such as patent awards or cost reductions in specific processes. The amount of the incentives can be determined beforehand as a specific dollar figure or vary as a percentage of financial results. Finally, incentives can be limited to the key executives of the company or apply to the majority of the employees of the firm.

Financial incentives can vary widely in form and structure, but of course all managers receive a base salary. Bonuses can be paid immediately, deferred over a number of years, or given in the form of stock. Options, rights, and warrants entitle their holders to buy shares in a company for a specific price guaranteed for a specific period. These holders stand to benefit from any increase in value of company shares over the years. Thus, the holders have a direct incentive to maximize the company's financial performance. Changes in the federal tax system over the years and in the stock market in general have also influenced the form of financial incentives. These issues are discussed in greater detail in chapter 6. Higher capital-gains-tax rates in the 1970-1978 period and more restrictive tax rules applicable to options, together with a decline of stock-market averages, have greatly reduced the attractiveness of stock options.

In general, incentives for patents and innovation are of three types. Federal rules specify that patent rights cannot be given away for free by individuals to their employer. But companies have many ways to reward innovative individuals. At the lowest level, companies may give a token $1 for each patent award. At a second level, companies may give a fixed dollar amount, first when a patent application is filed, and then when a patent is actually issued by the U.S. government. The amount given for filing a patent might be $50 to $100. When a patent is issued the reward might be around $500. This second type of reward for patent innovation does provide greater incentives for individuals to disclose their innovations to the company. The third type of award for patent innovations generally ties amounts given to either a percentage of cost savings or a percentage of revenues derived from the innovation. This percentage can vary among companies and can also be limited by a fixed-dollar maximum such as, for example, $25,000.

For many types of white-collar workers and managers, nonfinancial rewards are also important. Promotions within a company mean a more-prestigious title and scope of work, as well as salary increases. Usually, promotions will come more rapidly in a fast-growing company or in a company that experiences a large turnover rate. Therefore, individuals have every incentive to join a fast-growing company and then to ensure that its growth continues. The opportunity for young employees to move up rapidly in responsibility and salary may also have been an important factor in motivating individuals in the industry. This effect is obscured by the data presented earlier on average white-collar salaries because data are not available on the age distribution of employees. The same salary will mean more to a 25-year-old semiconductor employee than to a 35-year-old employee in another industry, for instance.

As a second major type of nonfinancial incentives, individuals may be allowed by their company to pursue independent research projects; companies may give them specific research budgets and provide them with research assistants. For individuals interested in advancing the state of the art or in developing new products and applications, this type of incentive can be very important.

Perquisites constitute the third type of nonfinancial incentive. The type and location of office space or the use of the company car are examples of perquisites.

Recognition by peers and superiors in the firm or outside the firm constitutes the fourth type of nonfinancial incentive. Recognition can take the form of special awards, write-ups in the company journal, appointments to or memberships in special committees, or permission to write articles for scientific journals and the trade press.

The Case of Intel

Within the semiconductor industry, the founders and employees of Intel benefited the most financially from the semiconductor revolution. As of the beginning of 1979, eight officers and directors of Intel each owned stock worth more than $1 million, with their combined holdings totaling about $200 million. The value of stockholdings is obtained by multiplying the number of shares owned by each individual, as described in Intel's proxy statement, and the market value of Intel shares, as reported in stock market quotations.

Even more remarkable is that over the past five years Intel granted and sold to its employees, other than directors and officers, options representing more than one and one-half times the combined holdings of these eight individuals. Roughly two-thirds of these options have been exercised by Intel employees. These employees represented close to 40 percent of the Intel workforce, or 3,800 employees. Intel's employee investments total roughly $60 million. In contrast, Intel has raised only $14 million from outside sources, including the original venture capital, according to an interview with an industry executive (Intel Corporation, proxy statement, 16 March 1979, p. 6).

Holders of stock options benefit financially when the market price of the shares, at purchase time or in later years, exceeds the exercise price of the option. Intel's market price was more than double the option price in early 1979, which means that Intel employees, other than the officers and directors referred to, have seen their net worth increase by more than $100 million through share ownership in Intel. This gain is equivalent to about $25,000 per participating employee.

In addition to the option plans mentioned here, Intel has granted options to thirty-three key employees to buy shares in a subsidiary set up specifically to develop a new technology, Intel Magnetics, Inc. (Intel Corporation, proxy statement, 16 March 1979, p. 6). Finally, as a special type of nonfinancial incentive, Intel gives selected outstanding researchers the freedom to pursue independent projects related to the business of the company.

Officer salaries and cash bonuses at Intel are generally in line (save for two exceptions mentioned later with those of other semiconductor firms and prevalent firms in the electrical-equipment industry. Compensation data for Intel and other companies are summarized in table B-1. Intel's proxy statement mentions that the general manager of the component group received $1 million in 1978 as extra remuneration, and that the general manager of the microcomputer-systems division received $2.1 million in cash bonuses from 1974 to 1978 under the management-incentives program (1979 proxy statement, pp. 5 and 14).

A remarkable feature of the Intel salary structure is that its range is limited for officers from slightly under $100,000 to slightly over $200,000. This spread is smaller than for other types of corporations and other semiconductor firms. Typically, in the electrical industry the average compensation ratios for the three highest-paid executives are 100, 68, and 54 (Top Executive Compensation, 1978, p. 45.) For Intel these ratios are 100, 91, and 80; among the semiconductor firms that we examined for 1977 and 1978, only National also presented such a narrow spread for top-executive compensation.

The example of Intel is unique among start-up companies in the semiconductor industry. No other start-up company has provided comparable financial returns to its founders or has rewarded its innovators and key employees with such lucrative stock options.

However, in and by themselves, the two major points—Intel's success and the financial benefits accruing to founders and employees—do not specify the sequence of events: first the employee incentives and then the success, or first the success and then the financial rewards. In fact, other companies appear to provide equally attractive salary and stock-option incentives, but since the companies have not been as successful, the stock options have not paid off the way Intel's have.

The only other instances in which major shareholders have benefited as much from the semiconductor revolution are those of two already established firms, Motorola and TI. These two companies and the newer companies, however, may have attempted to avoid the experience of Fairchild by providing adequate incentives. According to Braun and MacDonald (1978), "When companies have offered no stock options at all, they have sometimes seen their employees desert in consequence. Groups certainly left both Fairchild and Transitron for this reason in the days before those companies changed their attitudes toward stock options."

Incentives at Other Semiconductor Firms

This section examines three groups of semiconductor firms: large and broadly diversified firms, large and specialized firms, and finally, smaller firms.

Appendix B

Large and Broadly Diversified Firms

This category includes firms with semiconductor merchant sales such as Rockwell International, TRW, RCA, Harris, General Instrument, and Motorola. All these firms are vertically and horizontally integrated, and some of them have operations in many different lines of business. Therefore, semiconductor operations at most of these firms are but a minor part of total sales. Our analysis for this group will focus on Motorola, the diversified company with the largest position in the semiconductor industry.

The chairman of Motorola, Robert W. Galvin, and his family control directly or indirectly close to 10 percent of the Motorola stock. (This and other information was obtained from Motorola's proxy statement, 27 March 1979, p. 12*ff*.) At $50 per share, their net worth is about $150 million. No other director or officer at Motorola has shareholdings valued at more than $1 million: collectively, 86 Motorola officers (excluding the chairman) own stock worth between $6 and $7 million.

Motorola does have stock-option plans for its senior officers and key employees. These plans do not appear to be very popular with the employees since only a minor proportion of the options have been exercised. From November 1974 to March 1976, Motorola granted 1.2 million options to its employees under what it calls the "1972 stock-option plan": of these options only 56,660 were exercised by employees, which represents only a 5 percent exercise rate. All other options granted under this plan have expired since then. In contrast, Intel employees exercised two-thirds of the options granted to them so far, and part of these unexercised options are still outstanding.

Under a new plan started in 1976, Motorola granted 1.6 million options to its employees between the periods of March 1977 and January 1979. Of these options only 59,532 had been exercised by spring of 1979. Again, this amounts to less than 5 percent of outstanding options, although in this case some options have not expired yet.

No compensation data are available for the semiconductor division itself. But for the company as a whole, compensation for senior officers is in line with that of other semiconductor firms and the electrical-equipment industry (see figure B-2 and table B-1). For the five more-senior officers of Motorola, excluding the chairman, cash and noncash bonuses in fiscal 1978 represented between 60 and 80 percent of their basic salary. For the next 93 Motorola officers, bonuses amounted to 35 percent, on average, of base salary. This addition brought their average total compensation to about $84,000 per year.

Motorola also has a company-paid pension plan. Basically the plan states that, for senior officers, annual benefits will represent a maximum of 50 percent of their salary for the last year of employment. For other company employees, benefits will represent a maximum of 35 percent of average salary for the five best years, with a ten-year vesting period.

Large and Specialized Firms

The second group of companies consists of the large semiconductor firms including TI, National, and Fairchild. The distinction between these three companies and the first group has tended to blur over time, as TI and National have integrated forward into consumer products and commercial systems, and Fairchild has been acquired by a larger, diversified company. Officers and directors of TI hold, as a group, close to one million shares of the company. This amount represents about 4.3 percent of outstanding shares. At $100 per share, their combined holdings are worth about $100 million. The shares owned by TI's chairman are worth $8 million; shares owned by the president are worth slightly over $5 million. A former TI chairman, Patrick E. Haggerty, controls shares worth about $43 million. He is the one director or officer who owns the most shares.

As of the beginning of 1979, the five top officers held 324,000 options. Other TI officers held 460,000 options, while nonofficer employees held 508,000 options. The share-option price between officers and nonofficer employees differs markedly; the price for officers was slightly below $75, while the price for nonofficers was about $94 per share. Total compensation for TI senior officers in 1977 is comparable to that of the industry (figure B-2), but it is heavily weighted toward incentive payments. For 1978 compensation was divided about equally between base salary and incentive payments. For fifty-two other TI officers, 1978 compensation consisted of base salary for two-thirds of the total and incentive payment for one-third. For the five senior TI officers, incentive payments are divided in four categories of roughly equal weight: immediate cash payment, future cash payment, immediate equity payment, and future equity payment. For the other officers incentive payments are weighted more heavily on immediate payments.

National is very much an anomaly, even for the semiconductor industry, with regard to its incentive packages. National financial and nonfinancial incentives are definitely below industry averages, yet the company does not appear to have difficulty in attracting and retaining effective managers or in obtaining a good performance.

Stockholdings of company officers are among the smallest in the industry. National officers own about 3.7 percent of company shares. Compensation packages are noticeably below those available at companies of similar size. National's compensation for officers is roughly similar to that at companies one-fifth its size (see figure B-2).

Bonuses are also small in absolute amounts and in proportion to salary. For fiscal year 1978, bonuses did not exceed $24,000 for any officer, and on the average bonuses represented less than 25 percent of base salary for all officers (National Semiconductor, proxy statement, 9 August 1978).

Options available to National officers represent around 2 percent of outstanding shares. These options are exercisable at an average price of about $23 per share; market price of National shares in the fall of 1979 was around $30. The three top officers each own between 30,000 and 35,000 options, giving them a pretax paper profit of almost $200,000. Eleven other National officers own 14,000 options each on average.

Control of Fairchild Camera and Instrument Corporation was bought in 1979 by a diversified firm, Schlumberger Limited. Prior to that the largest single shareholder of Fairchild was the Sherman Fairchild Foundation, which controlled about 11.6 percent of the shares. Fairchild senior officers owned very few shares. Before the takeover, the then president of Fairchild, Wilfred J. Corrigan, owned less than 6,000 shares worth about $215,000 (Fairchild Camera and Instrument Corporation, proxy statement, 16 March 1979).

The past president, C. Lester Hogan, owned slightly more than 18,000 Fairchild shares, worth around $700,000, but 10,000 of these shares were pledged to Fairchild for a loan. Dr. Hogan had reduced his Fairchild holdings over the years. In 1968, he made the headlines in the semiconductor industry when he left his position as general manager of Motorola's Semiconductor Division to become president of Fairchild. As described by Braun and MacDonald (1978, p. 132), "He was offered a salary of $120,000, an interest-free loan of $5.4 million to exercise an option of 90,000 shares at $60 each, and a further allotment of 10,000 shares at $10 each." Financial packages offered to semiconductor executives during the following few years would be measured in terms of Hogan's package. Managers would say that they were offered a quarter or a half by Hogan to transfer companies (*Business Week*, 4 October 1969).

Compensation for Fairchild officers appears in figure B-2 and table B-1. Compensation packages of Fairchild officers are roughly one-half of those for TI officers, and save for the highest-paid position, about comparable to packages for National officers.

Smaller Semiconductor Firms

Smaller semiconductor firms, with annual sales at or below $100 million, constitute our third group of companies. The analysis will focus on four of them: AMD, AMI, Intersil, and Mostek.

Advanced Micro Devices (AMD) officers and directors own collectively 6.2 percent of company shares worth about $5.4 million at a share price of $28 (AMD, proxy statement, 12 September 1978). AMD's president Jerry Sanders owns 47,000 shares worth roughly $1.5 million. The largest single stockholder is a German electronics firm, Siemens, with close to 20 percent of AMD stock. Officers and directors of AMD own options equivalent to 4.7 percent of outstanding shares. Mr. Sanders owns half of these options and the senior vice-president for production owns about 10 percent of them.

Salaries and bonuses of AMD officers are comparable to those at other semiconductor firms (see figure B-2 and table B-1). From figures B-1 and B-2 apparently such compensation is somewhat higher than is prevalent in the electrical-equipment industry. However, no firm conclusion can be reached, since the conference board does not report on compensation for firms with sales below $100 million per year. Possibly compensation levels flatten out for firms with sales below, say, $150 million.

Bonuses at AMD are comparatively high, their range extending from 40 percent of base salary for lower-ranking officers to 80 percent for the president. AMD's document specifies that incentive compensation for AMD's officers are not to exceed 50 percent of base salary. This maximum was specifically waived for one senior officer. No regular stock-option plans are available for employees who are not officers of the company.

AMD's proxy statement is unusual in that it describes in detail the terms of employment of its president. Mr. Sanders has a contract that runs to March 1983, guaranteeing him a yearly salary of $133,000 plus an adjustment of 7 percent per year plus a cost-of-living allowance. His contract also specifies that he would get as incentive compensation 1 percent of AMD's profit before taxes for the fiscal year ending March 1979. Thereafter, Mr. Sanders is to get half of 1 percent of AMD's profit before taxes as incentive compensation. The contract further specifies that Mr. Sanders participate in the companywide incentive plan and that he may also get other compensation at the discretion of the board of directors. All payments made to Mr. Sanders under this contract are reflected in figure B-2 and table B-1.

In addition to compensation received directly from AMD, Mr. Sanders receives compensation from a European joint venture between AMD and Siemens. Mr. Sanders' contract with the joint venture specifies that he will receive $48,000 per year, plus 5 percent of profits before taxes of the joint venture, plus participation in the companywide incentive plan.

American Microsystems Inc. (AMI) management owned about 5.5 percent of company shares at the beginning of 1978 (American Microsystems, Inc., proxy statement, 5 April 1978). These shares were worth slightly over $2 million at that time and had doubled in value by the fall of 1979. Most of AMI's senior management was changed following financial problems at the company in the mid-1970s. Of the ten senior company officers, eight were

Appendix B

hired in 1976 or later, while the two others are not founders of the company and together own less than 1 percent of outstanding shares. All of these officers have had long experience with other firms in the industry.

The largest AMI shareholder, with a 25-percent share, is a company jointly owned by Borg-Warner and a German firm, Robert Bosch (BW-RB). These shares were acquired directly from AMI in June 1977. AMI by-laws and an agreement between AMI and BW-RB effectively restrict BW-RB's influence in AMI affairs. The BW-RB group is limited to a board representation proportional to its stock interest. Of the other board members only one, the president, is an officer of the company; the other members are outsiders.

AMI's proxy statement shows that every effort was made to interest senior management in the profitability of the company. AMI lent close to $1.5 million to the three most-senior officers to enable them to buy the shares mentioned in the previous paragraph. It also granted them options and warrants roughly equivalent in number to the shares they already owned.

Compensation for the AMI officers in the year 1977 is comparable to that at other semiconductor firms. Bonuses are roughly equivalent to 50 percent of base salary for the president and 25 percent of salary for the other officers.

AMI's bonus plan for key executives specifies that the pool available for bonuses in 1978 was to be equivalent to between 6 and 12 percent of profit before taxes, if the 1978 goal for profits was attained. If the 1978 goal was exceeded, the bonuses were to be augmented by between 10 and 20 percent of the extra profits. If the 1978 goal was not attained, bonuses were to be reduced accordingly. The plan also specified that bonuses were to be paid 100 percent in cash, but that if such bonuses were used totally for acquiring stock options, an extra 20-percent bonus was to be paid. This method is a further incentive for professional managers to link their own financial future to that of AMI.

Again, in the case of Intersil, management owns only a small proportion of the outstanding shares. A Canadian firm, Northern Telecom, owns 22.8 percent of Intersil (Intersil Inc., proxy statement, 11 December 1978). At $16 per share, this block of stock is worth slightly more than $20 million. Over the last five years, more than 700,000 options were granted to Intersil employees at an average exercise price of $6; market price was $16 in mid-1979.

Salaries and bonuses for Intersil officers are in line with those of the industry (figure B-2 and table B-1). Intersil's proxy statement describes how bonuses are determined for officers of the company other than the president. All officers share in 10 percent of Intersil's profit before taxes. Key officers share in a further 2.5 percent of profits before taxes.

Compensation of Mostek Corporation officers is comparable to that of other semiconductor firms (figure B-2 and table B-1); the company does not break down salaries and bonuses.

Mostek agreed during 1979 to be acquired by United Technologies Corporation (UTC) after having refused a takeover bid by Gould (*Wall Street Journal*, 28 September 1979). Mostek had also been approached before by foreign manufacturers interested in acquiring an ownership position but had always declined to make any commitment. At the time of its acquisition Mostek management owned 5.9 percent of the company, and a block of 20.8 percent was owned by Sprague Electric (Mostek Corporation, proxy statement, 20 April 1979). The price offered by UTC, $62 per share, was more than triple the market price at the beginning of the year. Proceeds to Mostek management would be in excess of $17 million, with about $6 million going to the president, Mr. Sevin. Ironically, each one of the five senior Mostek officers had let 20,000 options expire by the end of 1978 at an average exercise price of $14. If they had been exercised, such options would have given Mostek management an extra gain of $5 million when the UTC bid was announced.

Summary

This appendix has discussed incentives within firms in the semiconductor industry and presented data on salaries and stock-option incentives. Nearly all firms provide incentives for key employees based on corporate performance. The Intel example illustrates the difficulty of distinguishing cause and effect. Incentives encourage individual effort, but the profits accompanying successful firm performance allow employees to exercise incentives such as stock options. Except for Intel, the rewards achieved by individuals are similar across companies. Furthermore, salaries and incentive compensation in the semiconductor industry are in line with averages for other manufacturing industries.

Bibliography

Allan, Roger. 1979. "VSLI: Scoping Its Future." *IEEE Spectrum*, 16, no. 4: 31-3

Altman, Lawrence. 1977. "Special Report, MEMORIES, It's a User's Paradise: Cheaper RAMs, Reprogrammable ROMs, CCDs and Bubbles Coming Along." *Electronics*, 20 January, pp. 81-96.

———. 1976. "Advances in Design and New Process Yield Surprising Performance." *Electronics*, 1 April, pp. 74-81.

———. 1972. "Special Report: Semiconductor RAMs Land Computer Mainframe Jobs." *Electronics*, 28 August, pp. 63-77.

American Enterprise Institute for Public Policy Research (AEI). 1978. "The Aministration's 1978 Tax Package." Legislative Analyses, no. 28, 95th Congress, 15 May.

Arrow, Kenneth J. 1962. "Economic Welfare and the Allocation of Resources for Invention." In *The Rate and Direction of Incentive Activity: Economic and Social Factors*, Richard R. Nelson, Princeton, N.J.: Princeton University Press.

Asher, Norman J., and Strom, Leland D. 1977. *The Role of the Department of Defense in the Development of Integrated Circuits.* Prepared for the Office of the Director of Defense Research and Engineering, Institute for Defense Analyses, Program Analysis Division, IDA Paper P-1271, May.

Bloch, Erich. 1979. Direct Testimony in Defendent's Exhibit 9115A. *U.S. v. IBM*, transcript, 9 October, pp. 91466-91565.

Bullis, W.M. 1978. "Government Programs on Advanced Technology and Manufacturing Techniques: Comments on U.S.A., Japan, and Europe." Prepared for National Bureau of Standards, Washington, D.C., June.

Braun, Ernest, and MacDonald, Stuart. 1978. *Revolution in Miniature.* Cambridge: University Press.

Business Week. 1979. "American Manufacturers Strive for Quality—Japanese Style." 12 March, pp. 32C-32T.

———. 1979. "The Super Growth in Memory Chips." 3 September, pp. 121, 125, 128.

———. 1979. "Boom Times at Last at Mostek." 10 September, pp. 86, 90.

———. 1979. "Mostek and the Vanishing Pioneers." 15 October, p. 138.

———. 1979. "Can Semiconductors Survive Big Business?" 3 December, pp. 66-69, 77.

———. 1979. "Japan is Here to Stay." 3 December, pp. 81, 85-86.

_____. 1978. "Texas Instruments Shows U.S. Business How to Survive in the 1980s." 18 September, pp. 65, 76, 81.

_____. 1978. "The Pentagon's Push for Superfast ICs." 27 November, pp. 136-142.

_____. 1977. "Fairchild Problems: More than Watches." 15 August. pp. 116-117.

_____. 1976. "How to Survive in Semiconductors." 22 November, pp. 66D-66S.

_____. 1974. "The Semiconductor Becomes a New Marketing Force." 24 August, pp. 34-42.

_____. 1971. "Cashing in on a New Generation of Chips." 27 March, pp. 50-52.

_____. 1971. "An Integrated Circuit that is Catching Up." 25 April, pp. 134, 136.

_____. 1970. "Why Sylvania's Out of Semiconductors." 10 October, p. 29.

_____. 1969. "Signetics Shows How in Semiconductors." 26 July, pp. 41-42.

_____. 1969. "Where the Action is in Electronics," special report, 4 October, pp. 86-98.

Bylinsky, Gene. 1973. "How Intel Won Its Bet on Memory Chips." *Fortune*, November, pp. 142-147, 184, 186.

Chang, Y.S. 1972. *The Analysis of the Offshore Activities of the Japanese Electronics Industry*. Washington, D.C.: International Bank for Reconstruction and Development, November.

Charles River Associates. 1980. *Innovation, Competition, and Government Policy: A Framework for Analysis*. Boston, Mass.

_____. 1979. *International Technological Competitiveness: Television Receivers and Semiconductors*, draft report. Prepared for the National Science Foundation, Boston, Mass.

_____. 1976. *Analysis of Venture Capital Market Imperfection*. Prepared for the National Bureau of Standards, Cambridge, Mass.

Dalton, James A., and Levin, S.L. 1977. "Market Power: Concentration and Market Share." *Industrial Organization Review*, 5: 27-36.

Datamation. 1974. "Microprocessor and Microcomputer Survey." December, pp. 90-101.

Dinnean, G.P., and Frick, F.C., 1977. "Electronics and National Defense: A Case Study." *Science* 165 (18 March): 1151-1155.

The Economist. 1979. "Microprocessors: A Bigger Bite." 6 October, p. 94.

EDP Weekly. 1979. "Financing Research and Growth: Top Priority for SIA in 1980s." 13 August, p. 2.

Electronic News. 1979. "House Panel Kills All Funds for DOD's VHSI Program." 14 May.

Bibliography

_____. 1979. 17 December, pp. 50-51.
_____. 1976. "Microprocessor." 8 March, pp. 1-58.
_____. 1975. "Microprocessors Feel Impact of Crowding Market." 13 October, pp. 1-34.
Electronics. 1980. "U.S. Markets Forecast 1980." 3 January. pp. 134-137.
_____. 1979. "U.S. Markets Forecast 1979." 4 January, pp. 114-117.
_____. 1979. "Italians to make Zilog Z8000." 15 March, p. 56.
_____. 1979. "Fairchild's Situation Attracts Takeover." 10 May, p. 46.
_____. 1979. "Mostek and the Vanishing Pioneers." 10 May, p. 138.
_____. 1979. "Venture Capital Booming Again." 10 May, pp. 88-89.
_____. 1978. "U.S. Markets Forecast 1978." 5 January, pp. 134-137.
_____. 1978. "8086 Microcomputer Bridges the Gap Between 8- and 16-bit Designers." 16 February, pp. 99-104.
_____. 1978. "National Aims at Low-End Units." 3 August, p. 50.
_____. 1978. "How Bit-Slice Families Compare: Part 1, Evaluating Processor Elements." 3 August, pp. 91-94.
_____. 1978. "Pentagon to Fund Major IC Program." 14 September, pp. 81-82.
_____. 1978. "VHSI Proposal Finds Willing Audience." 28 September, pp. 89-90.
_____. 1978. "Mostek to Make Intel's 8086." 23 November, pp. 46-48.
_____. 1977. "U.S. Markets Forecast 1977." 6 January, pp. 90-92.
_____. 1977. "Siemens, AMD Form Company to Make Microcomputer Systems." 13 October, pp. 31-32.
_____. 1976. "U.S. Markets Forecast 1976." 8 January, pp. 92-93.
_____. 1976. "19.95 Watch Coming from TI." 22 January, pp. 44, 46.
_____. 1976. "Calculators Take Offshore Trip." 5 February, p. 75.
_____. 1976. "Chronology—Significant Advances in Electronics Technology Reported over the Past Year." 28 October, pp. 138-139.
_____. 1976. "Motorola, Fairchild Technology Exchange Strengthens Intel's Microprocessor Rivals." 11 November, p. 30.
_____. 1975. "U.S. Markets Forecast 1975." 9 January, pp. 90-94.
_____. 1975. "Watch Market—Is 40 a Crowd?" 20 February, pp. 34, 36.
_____. 1975. "News Briefs." 6 March, p. 38.
_____. 1975. "Chronology—Significant Advances in Electronic Technology Reported over the Past Year." 16 October, pp. 26-27.
_____. 1975. "Recession Spurs Infringement Suits." 11 December, pp. 77-78.
_____. 1974. "U.S. Markets Forecast 1974." 10 January, pp. 107-110.
_____. 1974. "Custom LSI Fades into the Background." 10 January, pp. 74-76.
_____. 1974. "Semiconductor Random-Access Memories." 13 June, pp. 108-110.

_____. 1974. "Calculator-Chip Business Slows." 31 October, pp. 46-59.
_____. 1973. "U.S. Markets 1973 Forecast." 4 January, pp. 93-96.
_____. 1973. "Microcomputers Muscle In." 1 March, pp. 63-64.
_____. 1971. "GI Moves R&D East, Says It's in MOS to Stay." 6 December, p. 44.
_____. 1971. "Hogan Completes Fairchild Alignment." 6 December, pp. 42, 44.
_____. 1970. "Semiconductor Memories at a Glance: What's Here Now, What's on the Way." 2 March, pp. 143-146.
_____. 1970. "RCA Looks for Place in the IC Sun." 28 September, pp. 40-41.
_____. 1970. "Sylvania Closes Semiconductor Division." 12 October, p. 46.
_____. 1969. "Autonetics, Act 2." 13 October, pp. 54-55.
_____. 1968. "The Prospects Are Solid." 8 January, pp. 109-112.
_____. 1968. "Musical Chairs." 19 August, pp. 45-47.
_____. 1968. "The Dust Settles." 2 September, pp. 40-41.
_____. 1968. "Hogan Takes Hold at Fairchild." 30 September, pp. 119-122.
_____. 1968. "MOS Memory to Sell for 10 Cents a Bit." 30 September, pp. 189-190.
_____. 1967. "Surge in Semiconductors." 9 January, pp. 130-134.
_____. 1967. "LSI: The Technologies Converge." 20 February, pp. 124-127.
_____. 1967. "Integrated Circuits in Action." 6 March, pp. 149-157.
_____. 1967. "The Swing to TTL Becomes a Stampede." 18 September, pp. 179-182.
_____. 1966. "IC Industry: Picture of Health." 8 August, pp. 114-119.
_____. 1966. "Integrated Circuits in Action." 14 November, pp. 128-135.
_____. 1965. "The Expanding Market." 14 October, pp. 96-98.
Finan, William F. 1975. *The International Transfer of Semiconductor Technology Through U.S.-Based Firms.* Working Paper No. 118. Washington, D.C.: National Bureau of Economic Research, December.
Financial Times. 1977. "Zilog Aims to Lead the Micro World." 13 June, p. 8.
Financial World. 1977. "Programmable Microcomputers: The Next Bonanza." 15 March, p. 16.
Forbes. 1979. "The Micro War Heats Up." 26 November, pp. 49-58.
Fortune. 1979. "Texas Instruments Wrestles with the Consumer Market." 3 December, pp. 50-57.
_____. 1975. "Here Comes the Second Computer Revolution." November, pp. 134-184.

French, J.C. "Improvement in the Precision of Measurement of Electrical Resistivity of Single Crystal Silicon: A Benefit-Cost Analysis." NBS-Election Devices Section, Report No. 807. Washington, D.C., 20 September 1967.

Gale, Bradley T. 1972. "Market Share and the Rate of Return." *The Review of Economics and Statistics*, 54, no. 4: 412-423.

Gold, Martin. 1979. "Patterns in Pricing: Straightening the Curve." *Electronic News*. 31 December, p. 12.

Golding, Anthony M. 1971. "Semiconductor Industry in Britain and the United States: A Case Study in Innovation, Growth, and the Diffusion of Technology." Unpublished Ph.D. dissertation, University of Sussex, England.

Griliches, Zvi. 1958. "Research Costs and Social Returns: Hybrid Corn and Related Innovations." *Journal of Political Economy*, October, pp. 419-431.

Gutmanis, Ivars. 1979. Statement before the U.S. International Trade Commission on behalf of the Electronics Industries Association of Japan. Investigation No. 332-102. 29 May.

Heaton, George R., J.r,; Holloman, J. Herbert; and Ashford, Nicholas A. 1978. *Government Involvement in the Innovative Process*. A Contractor's Report to the Office of Technology Assessment by the Center for Policy Alternatives, Massachusetts Institute of Technology. October.

Hogan, C. Lester. 1979. Statement at the Technology Workshop on Semiconductor Electronics. Palo Alto, California, 14 June.

_____. 1977. "Reflections on the Past and Thoughts about the Future of Semiconductor Technology." *Interface Age*, 2, no. 4: 24-36.

Hulick, Charles. 1979. "Federal Procurement and Industrial Innovation: A Study of ETIP Concepts and Strategy." 23 May. Washington D.C.: U.S. Department of Commerce, Experimental Technology Incentives Program.

IC Master. 1977. Garden City, N.Y.: United Technical Publications, Inc.

Japan Economic Journal. 1972. "TI Will Get Ownership." 4 January, p. 2.

_____. 1971. "Electronics Industry Plans Cartel for Types of Products Using IC." 14 December, p. 9.

_____. 1970. "Mini Calculator to Be Sold." 5 May, p. 10.

_____. 1970. "Casio, Fairchild Tie-up for LSIs." 3 November, p. 10.

Kamien, Morton I., and Schwartz, Nancy L. 1978. "Self-Financing of an R&D Project." *The American Economic Review*, 68, no. 3: 252-261

Kindleberger, Charles P. 1968. *International Economics*, 4th ed. Homewood, Ill.: Richard D. Irwin, Inc.

Kleiman, Herbert S. 1977. *The U.S. Government Role in the Integrated Circuit Innovation*. Final Report. Prepared by Battelle-Columbus Labs for OECD. 25 February.

———. 1966. "The Integrated Circuit: A Case Study of Product Innovation in the Electronics Industry." D.B.A. dissertation, George Washington University, Washington, D.C.

———. 1966. "A Case Study of Innovation." *Business Horizons*, Winter, pp. 63-70.

Kraus, Jerome. 1973. "An Economic Study of the U.S. Semiconductor Industry." Ph.D. dissertation, New School for Social Research. New York, New York.

Lochwing, David A. 1973. "From 'Silicon Gulch' Came Dazzling New Applications for Integrated Circuits." *Barron's*, 17 September, pp. 5, 12.

Mansfield, Edwin; Rapoport, John; Romeo, Anthony; Wagner, Samuel; and Beardsley, George. 1977. "Social and Private Rates of Return from Industrial Innovation." *Quarterly Journal of Economics*, May, p. 236.

Mick, Colin K. 1979. "Threats to the Market Preeminence and Performance of the U.S. Semiconductor Industry." Decision Information Services Ltd., Palo Alto. Calif., 18 June.

Moore, Gordon. 1979. "VLSI: Some Fundamental Changes." *IEEE Spectrum*, 16, no. 4: 30-37.

National Academy of Engineering. 1979. NAE Colloquium on Public Policy and Industrial Innovation. 5-6 December.

Nelson, Richard R., Peck, M.J., and Kalachek, E. 1967. *Technology, Economic Growth, and Public Policy*. Washington, D.C.: The Brookings Institution.

The New York Times. 1979. "Winning and Losing at Texas Instruments." 13 May, p. F-1.

Nesheim, John L. 1979. Statement before the U.S. Senate Subcommittee on Taxation and Debt Management on behalf of the Semiconductor Industry Association. Washington, D.C., 18 June.

Noyce, Robert. 1979. Testimony on behalf of the Semiconductor Industry Association before the U.S. International Trade Commission. San Francisco, Calif., 30 May, p. 6.

———. 1977. "Microelectronics." *Scientific American*, 237, no. 3: pp. 62-69.

Office of Technology Assessment, U.S. Patent and Trademark Office. 1979. *OTAF Special Reports, Technology Profiles: Semiconductor Systems and Applications*. Washington, D.C., July.

Organization for Economic Cooperation and Development (OECD). 1968. *Gaps in Technology: Electronic Components*. Paris.

Osborne and Associates, Inc. 1978. *An Introduction to Microcomputers*. Berkeley, Calif.

Pacific Projects, Ltd., Tokyo, Japan. 1977. *The Japanese Market for Elec-*

tronic Components and Assemblies. Prepared for U.S. Department of Commerce, Domestic and International Business Administration. NTIS DIB-78-02-506. Washington, D.C., October, pp. 33-35.

Porter, Michael E. 1976. *Interbrand Choice, Strategy, and Bilateral Market Power*. Cambridge, Mass.: Harvard University Press, 1976.

Riley, Wallace. 1973. "Special Report: Semiconductor Memories are Taking over Data-Storage Applications." *Electronics*, 2 August, pp. 75-90.

Roberts, Edward B. 1968. "Entrepreneurship and Technology. A Basic Study of Innovators. How to Keep and Capitalize on Their Talents." *Research Management*, 2, no. 4: 249-266.

Rosenbloom, Richard S. 1978. "Technological Innovation in Firms and Industries: An Assessment of the State of the Art." In *Technological Innovation: A Critical Review of Current Knowledge*, P. Kelley and M. Kransberg, eds. San Francisco, Calif.: San Francisco Press, Inc.

Scace, Robert I. 1979. "VLSI In Other Countries." Notes for lecture to be given at an American university short course, June.

Schnee, Jerome. 1978. "Government Programs and the Growth of High-Technology Industries." *Research Policy* 7: 2-24.

Schuyler, Peter. 1979. "The Challenge of the Circuits." *The New York Times*. 4 April, p. D4.

Sickman, Phillip. 1966. "In Electronics the Big Stakes Ride on Tiny Chips." *Fortune,* June, pp. 120*ff.*

Sideris, George. 1973. "AMS Brushes off Blows." *Electronics*, 30 August, pp. 74, 76.

Status. 1973, 1975, 1976, 1978, and 1979. Publication of Integrated Circuit Engineering Corporation, Scottsdale, Arizona.

Tanaka, William H. 1979. Statement before the U.S. International Trade Commission on behalf of the Electronics Industries Association of Japan. Investigation No. 332-102. 30 May.

Tassey, Gregory. 1979. "Reforming Securities Regulations and the Policy Research Process." NBS Working Paper, June.

Texas Instruments. 1972. Management Systems, Intercollegiate Case Clearing House, Case #9-172-054.

Tilton, J. 1971. *International Diffusion of Technology: The Case of Semiconductors*. Washington, D.C.: The Brookings Institution.

Top Executive Compensation. 1978. New York: The Conference Board.

U.S. Department of Commerce. 1979. *Report on the Semiconductor Industry*. Washington, D.C.: U.S. Government Printing Office, September, p. 33.

U.S. Department of Commerce. 1961. *Semiconductors: U.S. Production and Trade*. Washington, D.C.: U.S. Government Printing Office.

U.S. Department of Commerce, Bureau of Census. "Shipments of De-

fense-Oriented Industries." *Current Industrial Reports*, 1965-1977, Series MA-175.
U.S. Internal Revenue Code. Sections 55-58, 1201, 1222, 1348.
U.S. Senate. 1978. "Industrial Technology, Hearings Before the Committee on Commerce, Science, and Transportation," 95th Congress, 2nd Session, 30 October, Serial No. 95-138.
U.S. Tariff Schedule. 1978. Schedule 8. Part 1. Annotated.
Utterback, J. and Murray, A. 1977. *Influence of Defense Procurement and Sponsorship of Research and Development of the Civilian Electronics Industry*. Cambridge, Mass.: Massachusetts Institute of Technology, Center for Policy Alternatives, June.
Vacroux, Andre G. 1975. "Microcomputers." *Scientific American*, 232, no. 5: 32-40.
Walker, C.E. and Bloomfield, Mark. 1979. "How the Capital Gains Tax Fight was Won." *The Wharton Magazine*, Winter, pp. 34-40.
Wall Street Journal. 1979. "The Super Chip." 27 April, pp. 1, 30.
_____. 1979. "Mostek Sets Merger Accord." 28 September, p. 10.
_____. 1979. "Reduction of Tax on Capital Gains Spurs Investment." 31 October, p. 24.
_____. 1979. W.R. Grace advertisement. 10 December, p. 9.
Webbink, Douglas A. 1977. *The Semiconductor Industry: A Survey of Structure, Conduct, and Performance*. Staff Report to the Federal Trade Commission. January.
Weisberg, Leonard R. 1978. "DOD Directions in Electronic Device R&D." *Proceedings of the 1978 IEEE Conference on U.S. Technological Policy*, pp. 24-25.
Wells, Louis T. 1972. "International Trade: The Product Life Cycle Approach." In *The Product Life Cycle and International Trade*, Louis T. Wells, ed. Boston, Mass.: Harvard University Graduate School of Business Administration, Division of Research.
Willig, Robert D. 1976. "Consumer's Surplus Without Apology." *The American Economic Review*, September, pp. 589-597.
Wolff, Howard. 1973. "4096-bit RAMs Are on the Doorstep." *Electronics*, 12 April, pp. 75-77.
Zschau, Edwin V.W. 1978. Statement before the U.S. Senate Subcommittee on Taxation and Debt Management. Washington, D.C. 29 June. Palo Alto, California: American Electronics Associaton.

Index

Advanced Memory Systems, 41, 91-92, 104-105, 107, 136
Advanced Micro Devices (AMD), 42-44, 48, 58-59, 63, 78, 81-82, 96, 103, 135, 149, 167, 186-187, 195-196
Advertising and promotion campaigns, 91
Agencies, governmental, 2
Aggressiveness: Japanese, 68; in pricing, 82, 94, 100-101, 137
Air Force, Department of the, 4, 148, 151-153; contracts for, 152
Aktiengesellschaft, 42-44
Allan, Roger, 35
Altman, Lawrence, 91-93
Aluminum, use of, 28, 86
Amelco, Teledyne subsidiary, 47
American Enterprise Institute, 165
American Microsystems Inc. (AMI), 16, 23, 40, 52, 57, 65, 88-89, 92-93, 136, 186-187, 195-197
American Telephone and Telegraph Corporation (AT&T), 4, 156
Annealing for stress relief, definition of, 181
Antimony impurity atoms, 28
Antitrust: enforcement, 173; laws and regulations, 156-158, 170-171; policies, 3-4, 141, 155, 174
Apollo Program, 149-150, 152
Arizona State University, 155
Armin Watch Company, 100
Army, Department of the, 4
Army Signal Corps, 151, 153
Arrow, Kenneth, 65-66
Asia, 159; Southeast, 34
Assembly: areas, 33; plants, 80, 161, 163
Associations, trade, 6
Availability, capital, 37-38, 46, 56, 62, 69, 135, 138, 142, 144, 152, 158, 164, 166, 168

Banking houses, 167
Bankruptcy proceedings, 100, 104
"Bare pipe-racks" image, 48
Barriers: to entry, 138, 142; mobility, 102, 112, 118, 142; nontariff, 161; to trade, 111; transistor, 47
Basic research, 42, 46, 62-68, 74, 155, 170
Batch-production process, 13
Battery drain, problem of, 29
Behavior: competitive, 6-7, 142, 156-157; innovative, 2, 5-6, 8, 25, 38, 46, 50, 69, 176
Bell Laboratories, 4, 13, 16, 41-44, 64-65, 154, 156-157
Benefits, social, 122
Benrus Watch Company, 100
Berkeley University, 155
Binary language, 28
Bipolar: circuits, 70; digital, 112-113; growth, 116; logic, 8-9, 77, 83-89, 104, 107, 120, 126; memories, 107; MOS combination, 180; shares, 114; technology, 16-17, 90, 92, 94-95, 104, 115; transistors, 28-29
Bipolar Junction Field-Effect (JFET), combination, 48, 180
Bloomfield, Mark, 168
Bobb, Howard, 52
Boeing Aircraft Corporation, 81
Bonders production equipment, 34
Bonding: labor-intensive process, 32-33; quality of, 78; ultrasonic, 181-182; wire, 163
Bonuses; cash, 191; noncash, 193; policy of, 169, 189, 195-197
Borg-Warner Corporation, 136, 197
Boron impurity atoms, 28, 31
Bosch, Robert, 136, 197
Bowmar Company, 99-100
Braun, Ernest, 13-15, 146, 148-149,

207

Braun, Ernest, cont'd.
153, 156, 170-171, 185
Bucy, Fred, 48
Budgets, cutbacks in, 154
Business, 50; cycles in, 11, 18, 166
Business Week, cited, 18, 21, 52, 54, 79, 87-88, 94, 99, 101, 129, 134, 136, 143, 145, 163-164, 167, 195
Buyers and sellers, 6
Bylinsky, Gene, 91-92

Calculators, 17, 19, 88, 98; desk, 28; electronic, 124; digital, 18; integrated circuits, 89; Japanese, 99-100; standard, 90
Canada, 92, 97
Canon Corporation, 99
Capital: advantages, 111; availability, 37-38, 46, 56-62, 69, 135, 138, 142, 144, 152, 158, 164, 166, 168; costs, 25, 130, 167; debt, 133; equipment, 32; formation, 166, 178; gains tax, 5, 73-74, 164-169, 174, 190; intensity, 104; investments, 11, 174; markets, 1, 17-18; requirements, 18; sources, 1; speculative, 17; stock, 165; venture, 1, 17-18, 58, 142, 164, 166-168, 174
Captive: production, 21, 80-81; suppliers, 25
Cash: bonuses, 191; flow, 178
Casio Company, 99
Ceiling-price time path, factor of, 35
Charles River Associates (CRA), 5, 8, 37, 134, 164, 169, 177
Chemistry and chemists, 13
Circuits: bipolar, 70, 112-113; consumer, 77, 83, 90, 120; custom, 87, 107; density, 1, 18; design innovations, 39; digital-integrated, 2-3, 7-8, 13, 18-19, 26-30, 45, 83, 88, 94, 100, 106, 158, 179; LSI, 22, 107; planar, 47; specialty, 19; technology, 13, 94; TTL, 119
Clevite Company, 23
CMOS, 52, 54, 153; logic, 119; silicon-on-sapphire technology, 93; transister, 39, 180
Cogar Corporation, 91, 171
Color television receiver industry, 134
Commercial: applications, 2; computers, 152; data processing, 16; innovation, 175; objectives, 64; technology, 143
Commodore Company, 99-100
Communications, 152; equipment, 189; satellite, 143
Competition, 144, 173; appraisal of, 83; behavior, 6-7, 142, 156-157; cost of, 129; forces of, 89; foreign, 1-2, 4, 6, 11, 21-22, 25, 68, 134, 159, 163, 178; intense, 67; international, 157, 175; Japanese, 142, 156-157, 164; market, 26; performance, 9; pressure, 26; price, 77, 92, 100-104, 106, 137; strategies in, 2, 5, 8, 50, 177
Completentary Metal Oxide Semiconductor Transisters. *See* CMOS
Computer(s), 18-19, 21, 24, 28, 90; commercial, 152; designs, 29, 34-35, 87-88; electronic, 91; equipment, 66, 127; logic, 34; manufacturers, 80; memory, 17; microtechnology, 91; small, 86
Conglomerates, 4
Congress, actions of, 165-166, 171
Consent Decree of 1956, 156
Constraints: legal and societal, 6; price, 26
Consultants, industrial, 141
Consumer(s): circuits, 77, 83, 90, 120; markets, 78; products, 9, 25, 80, 98-101; surplus, 122-130; users, 18-19, 23
Consumer Electronics Show, 25, 100
Contract(s): Air Force, 152; cost-sharing, 152; funding terms, 78; Minuteman II program, 148-149; procurement, 2, 18, 74, 147-148, 150
Controls: electronic, 47; management, 37, 45-46, 50-55, 69, 72
Cornell University, 155

Index

Corning Glass Corporation, 167
Corrigan, Wilfred J., 195
Cost(s): capital, 25, 130, 167; competitive, 129; consciousness, 150; electronic functions, 13; entry, 167; fixed, 6; integrated circuit, 26; labor, 163; living, 196; manufacturing, 34; patent filings, 63; production, 11, 30, 34, 48; research, 64; sharing, 152
Corporate: goals, 6, 8, 175; management, 51; performance, 67; profitability, 112, 116; strategy, 81
Countercyclical devices, 165
Court cases and rulings, 170-171
Credit, tax, 164-165, 170-171
Cross-licensing agreements, 62-63, 156
Crystal growing, 30-31, 34
Custom: circuit, 87, 107; design, 78; LSIs, 35, 77, 83, 86-90, 107, 136; MOS, 88
Customer: acceptance, 94; contact, 66-67; markets, 6; military, 93
Cycles, business, 11, 18, 166

Dalton, James A., 114
Data processing, 16
Datamation, 95, 121-122
Dataquest, 113, 115, 119, 130
Debts and debtors, 133
Decentralization, policy of, 73
Decisions: innovative, 58, 60-61; investment, 51; process of, 59
Deep Ultraviolet Photolithography, definition of, 183
Defense, Department of (DOD), 3, 143, 146, 150-151, 155, 163, 170
Delivery commitments, 82, 94
Demand factor, 2, 8, 26
Demography, changes in, 12
Density, circuit, 1, 18
Depletion-mode MOS Resistor, definition of, 180
Deposited metal resistor, definition of, 180
Designs: circuit innovations, 39; computer, 29, 34-35, 87-88; custom, 78; logic, 85
Desk calculators, 28
Dielectric isolation, definition of, 182
Diffusion, factor of, 16, 179
Digital: bipolar, 70, 112-113; calculators, 18; logic circuits, 19; watches, 19, 100
Digital integrated circuits, 11, 62, 158; concept, 28-30; demand, 83; industry, 106; innovation, 68; product areas, 19, 88; technology, 94; transistor roots, 26-29
Diode-Transistor Logic (DTL), 30, 84-87, 184
Direct Bonding (Bump), definition of, 182
Direct-Coupled Transistor Logic (DCTL), definition of, 84
Discretionary Wiring techniques, 87-89, 107
Discrimination, price, 157
Distribution network, 25, 77-80
Diversity of strategies, importance of, 120-122
Double-polysilicon technology, 93
"Dumping" semiconductors, 157-158
Dynamic-memory field, 94

Eastman Kodak Company, 80
Economist, The, cited, 97
Economy and economic policies, 6, 30, 142, 165-166
EDP Weekly, 158
Education: factor of, 6; salesperson's, 66
Electrical industry, 48, 192
Electron-Beam Processing, definition of, 183
Electronic Arrays, 52, 81, 88, 93, 100, 136
Electronic News, cited, 5, 96, 144
Electronics, cited, 52-53, 84-85, 87-91, 93, 95-98, 100, 126, 135, 143, 167-168

Electronics: applications, 134; calculator, 124; components, 150, 153; computers, 25, 91, 100; consumer products, 25; controls, 47; costs per function, 13; equipment, 21, 80; products, 24-25, 99; research in, 155; solid-state, 170; warfare, 143
Emitter-Coupled Logic (ECL), 47, 184
Employees: and ownership plans, 168; theft by, 171; white-collar, 67
End users, factor of, 25
Energy, Department of, 155
Enforcement policies, 173
Engineers and engineering, 13, 47, 67, 109; electrical, 48; incentives, 168; intensive, 144; production, 60; programs, 171; resources, 83; reverse, 102; skilled, 33, 174; supply of, 5
Entrepreneurs: ability of, 35; incentives for, 1; and technology, 46-47, 74
Entry, 173; barriers to, 138, 142; capital costs of, 167; conditions of, 6; into industry, 11; patterns of, 13, 16-18
Environment: industrial, 6-7, 11-16; for innovation, 173
EPA regulatory policies, 5
Epitaxial techniques, 16
Equipment, 165; capital, 32; communications, 189; computer, 127; electronic, 21, 80; industrial, 18, 86; innovations, 33; manufacturers, 25, 80-81, 87, 89; plant, 6; production, 34; specialized, 54; telecommunications, 156
Erasable Programmable Real-Only Memories (EPROM), definition of, 19, 184
Europe, 22-23, 143, 155, 159; sales in, 131-133, 137
Executives, industrial, 2, 8, 37, 44, 46-49, 54, 58-59, 62-63, 66-67, 71-73, 79, 82, 141, 143-144, 152-155, 158, 169-171, 174

Exemptions, tax, 169
Expenditures, research and development, 5, 164-165
Experience curve, role of yields, 6, 11, 26, 32-34
Exports, factor of, 3-4, 159-160, 162; markets, 169-170
Exxon Corporation, 98

Fabrication technology devices, 143
Fairchild Camera and Instrument Corporation, 16-17, 21, 23-24, 35, 40, 42-44, 48, 51, 53, 55-58, 63-65, 70, 85-89, 92-94, 96-97, 99, 101, 103-105, 107-109, 135-136, 144, 149, 152, 167, 171, 177, 186-187
Fantasia Calculator Corporation, 100
Federal policies, 1; tax system, 170-171, 190
Feedback: performance, 111, 135-137; self-reinforcing, 16; strategic, 138
Finan, William F., 64, 105, 159, 161
Financial: incentives, 37, 46, 62, 67-68, 138, 185, 189; performance, 111, 135; resources, 1, 6, 38; rewards, 185, 189-190, 192; risks, 57-58
Financial Times, cited, 98
Financial World, cited, 95, 97, 122
Fixed costs, 6
Flexibility, organizational, 37, 45-46, 50-55, 69
Forbes, cited, 163, 167
Ford Motor Company, 51
Foreign: competition, 1-2, 4, 6, 11, 21-22, 25, 68, 134, 159, 163, 178; earned income, 5; firms and companies, 104-105, 138, 142, 145, 157-158, 174; governments, 133, 155; manufacturers, 1; markets, 178; ownership, 145; tariffs, 161; technology, 145; trade, 163
Fortune, cited, 25, 95-96, 122
France, 1, 24, 161
Freedom of Information Act, 155
Free-world countries, 159
Fujitsu Company, 24, 93-94

Index

Fund advanced development programs, 52
Funds and funding: contracts, 78; governmental, 3, 5, 133, 151-155, 170, 172-173, 177-178; procurement, 154; research and development, 3-5, 18, 142, 144, 151-155, 172-173, 177-178

Gale, Bradley T., 114
Gallium arsenide, 152
Galvin, Robert W., 193
Games: children's, 18; video, 19, 28
General Electric Corporation, 22-24, 42-44, 51, 53, 55, 65, 68, 137, 151, 156-157
General Instrument Corporation, 23 25, 41-44, 57, 65, 88, 95, 97, 99-100, 108, 193
General Microelectronics (GMe), 17, 40, 65, 82, 88, 136
General Telephone and Electronics Corporation, 52
Germanium-mesa transistor, 13, 147
Germany, 1, 24, 132, 161. *See also* West Germany
Gettering processes, definition of, 181
Gillette Watch Company, 100
Glass-chemical vapor deposition, definition of, 182
Goals, corporation, 6, 8, 175
Gold, Martin, 119
Gold-bonded diode, 16, 148
Golding, Anthony M., 51, 84, 86, 135, 146-149, 152, 156
Good will and goods and services, 3, 6
Gould Company, 198
Government: agencies, 2; foreign, 133, 155; funding, 3, 5, 133, 151-155, 170, 172-173, 177-178; intervention, 178; policies, 141; regulations, 5; research, 4; tax, 2; university support, 5
Grants, awarding of, 171
Great Britain, 163

Griliches, vi, 122
Grove, Andrew, 47
Gruen Watch Company, 100
Gutmanis, Ivars, 143, 147, 149

Haggerty, Patrick E., 47, 194
Hardware, military, 84
Harris Company, 41, 54, 96, 193
Harwood, Charles, 145
Heat, dissipation of, 95
Hewlett-Packard Corporation, 25, 80, 99
High-technology industries, 171
Hitachi Corporation, 24, 42-44, 64, 93-94
Hittinger, William, 52
HMW Industries, 100
Hoerni, Jean, 47, 171
Hogan, C. Lester, 2, 35, 47, 135, 154-155, 171, 195
Holdings, stock, 67
Home markets, 157
Honeywell Corporation, 42-44
Howe, David, 77
Hughes Corporation, 16, 23

IC Master, cited, 95-96
Illinois, University of, 152
Import: competition, 1; policies, 3-4
Incentive(s): compensation, 67, 196, 198; engineering, 168; entrepreneurial, 1; financial, 37, 46, 62, 67-68, 138, 185, 189; innovative, 7; investment, 164-165; nonfinancial, 190-191; systems, 2, 24, 46
Income: earned, 170; foreign, 5; overseas, 171-172; personal service, 166, 169; worldwide, 170
Increasing Function Density, definition of, 34-35
Incremental innovation strategies, 116-118, 134, 137-138, 142, 176-178
Industry: consultants, 141; digital integrated circuits, 106; electrical, 192; entry into, 11; environment, 6-7, 11-16, 23; equipment, 18-19,

Industry: cont'd.
86; high technology, 171; markets, 147
Information retrieval and storage, 90-91, 122
"INN2," factor of, 70, 72
Innovation: behavior, 2, 5-6, 8, 25, 38, 46, 50, 69, 176; characteristics, 43-46, 69-72; commercial, 175; decisions, 58, 60-61; digital integrated circuit, 68; encouragement, 67-68; environment, 173; equipment, 33; incentives, 7; incremental strategies, 116-118, 134, 137-138, 142, 176-178; market, 89; opportunities, 11, 13; pressures, 167; rents, 131
Integrated circuits: calculator, 89; costs, 26; development, 23; digital, 68, 106; industry, 106; MOS, 65; planar, 47; manufacturing process, 30-32; proprietary, 84
Integration, vertical, 11, 21, 23, 25, 77, 80-81
Intel Corporation, 16, 21-25, 40-44, 47, 50-53, 55, 57, 63, 68, 82, 88-100, 103-105, 107, 121, 130, 136, 143, 167, 185-187, 191-192, 198
Intensity: capital, 104; and competition, 67; engineering, 144; labor, 54
Interdependency policies, 7, 141, 172-174
Interest rates, 73
Interface Company, 54
Interfirm mobility and interactions, 141, 170
International: competition, 157, 175; performance, 111, 118, 131-134, 142; trade rules, 142, 158, 163
International Business Machines (IBM), 22, 24-25, 40, 44, 64-65, 68, 80, 171
International Revenue Code (IRC), 166, 168
International Telephone and Telegraph (ITT), 24
Intersil Corporation, 47, 96, 136, 171, 186-187, 195, 197

Intervention, government, 178
Interviews, results of, 45-46, 58, 64, 66-67, 71, 79, 82, 141, 143-144, 152-155, 158, 169-171, 173-174
Inventions, 70
Investment Tax Credit (ITC), 4-5, 164-165
Investments: capital, 11, 174; decision making on, 51; incentives for, 164-165; levels of, 61
Ion Implantation and Milling, 182-183
Isoplanar, coplamos, definition of, 180
Italy, 24

Japan, 1, 22-24, 93-94, 136, 159, 161; aggressiveness, 68; calculators in, 99-100; competition in, 142, 155-157, 163-164; conglomerates, 4; firms, 63-64, 79, 105, 134; markets, 157-159; producers, 106; sales, 131-133
Japan Economic Journal, cited, 99, 105, 132
Justice, Department of, 62, 156, 158, 173

Kamien, Morton I., 56
Kindleberger, Charles P., 157
Kleiman, Herbert S., 4, 151, 154, 173
Korea, 34, 163
Kraus, Jerome, 17-18, 47, 52, 65, 136, 156

Labor: costs, 163; forces, 17; intensity, 54, 161, 163; mobility, 178, skilled, 5, 54, 174
Laissez faire system, 1-2, 118, 133, 138, 141
Language, binary, 28
Large-scale integrated (LSI) memories, 8-9; bipolar, 89; circuits, 22, 107; custom, 35, 77, 83, 86-90, 107, 120, 136; market, 150; standard, 88
Laws: antitrust, 156-158; taxes, 5, 169-170
Leadership styles, 6, 94
Legal: barriers to entry, 138; restrictions, 73; societal constraints, 6

Index

Less-developed countries, 161
Levin, S.I., 114
Licenses and licensing: agreements, 26; cross, 62-63, 156, 173; liberal, 18; low, 173; marketing, 64; patent practices, 5, 145, 156-157
Lieberman, Marvin, 77
Life styles and cycles, 11, 77, 104
Light-emitting diode (LED), 101
Linear integrated-circuit technology, 19, 48
Literature, trade, 8, 46
Lithography, factor of, 143
Litrouix Watch Company, 100
Living allowances, cost of, 196
Logic: bipolar, 8-9, 77, 83-87, 104, 107, 120, 126; circuit category, 19; CMOS, 119; computer, 34; designers choice, 85; emitter-coupled, 47, 184; families, 184
Low-power Schottky, 54

MacDonald, Stuart, 13-15, 146, 148-149, 153, 156-157, 170-171, 192, 195
Mackintosh, I.M., 23
Magnetic: cores, 91; stereo tape, 90
Magnetics (EMM), 96
Malaysia, 163
Management and managers, 67-68; control, 37, 45-46, 50-55, 69, 72; corporate, 51; resources, 79; techniques, 46-50; top, 38, 46-50, 70-73, 133, 169
Manpower, 3, 5; mobility, 171; philosophy and policies, 6, 30-32, 170-172; skilled, 60, 171-172
Mansfield, Edwin, 122
Manufacturers: computer, 80; costs, 34; equipment, 25, 80-81, 87, 89; foreign, 1; semiconductor, 1
Markets and marketing, 45, 77; applications, 98; capital, 1, 17-18; competition, 26; consumer, 78; customer, 6; distribution, 79-80; end-product, 25; export, 169-170; foreign, 178; home, 157; industrial, 147; innovations, 89; input, 6; internal, 109;

Japanese, 157-159; licenses, 64; LSI, 150; merchant, 157; new, 48, 147; opportunities, 91, 147; places, 83, 124; power, 5; risks, 38, 60-61, 74; staffs, 37-38, 66-67; strategies, 82; structure, 6, 156, 177; target shares, 6; worldwide, 111, 147, 164
Massachusetts Institute of Technology (MIT), 152, 155
Matsushita Company, 24, 64, 96
Memory systems and memories, 77, 88-89, 120; advanced, 41, 91-92, 104-105, 107, 136; bipolar, 107; computer, 17; dynamic field, 94, 122, 124; large-scale integrated, 8-9, 83; monolithic, 96; race, 90-94; random-access, 19; read-only, 19; semiconductor, 91; 16K, 26; storage data, 19, 34; technology, 92, 94, 103
Merchant sales, 21-22, 144
Merged transistor, 181
Mergers, policy on, 135-138, 144
Metallurgy and metallurgists, 13
Metal-Oxide Semiconductor (MOS), 28-29, 39, 165; capacitor, 179; combination bipolar, 180; custom, 88; dynamic, 94, 122, 124, 126, 130, 138; integrated circuits, 65; memories, 92, 94, 103, 122, 124; resistor, 180; revenues, 114; sales growth, 88, 116; technology, 16-17, 70, 90-91, 95, 100, 104-105, 115, 136, 153; transistor, 179
Mexico, 163
Meyer, Leslie, 11
Mick, Colin K., 168
Microeconomic effects, 5, 8-9, 19, 29, 48, 66-67, 77, 83, 90, 95-98, 103, 105, 109, 114, 119-120, 135-136, 184
Microelectronic reliability program, 153, 163
Micromatrix, 87-89, 107
Microsystems International Ltd., 92, 96-97
Microtechnology, computer, 91
Microwave detection, 13
Military, 19, 23; business, 148; cus-

Military, cont'd.
 tomers, 93; production, 147; specification devices, 78, 143, 149; technology, 143; use output, 3, 175
Minority interests, 132
Minuteman Missile Project, 84, 147-149
Mobility: barriers, 102, 112, 118, 142; interfirm, 141, 170; labor, 178; manpower, 171; personnel, 73, 103, 138; skilled personnel, 17
Moore, Gordon, 35, 47, 185
Moore's Law, 34-35
Mostek Corporation, 16, 25, 40-44, 55, 57, 91-97, 100, 107, 130, 135-136, 144-145, 167, 186-187, 195, 198
Motorola Corporation, 16-17, 21, 23-25, 41-44, 47, 53, 57, 63, 85, 88, 91-98, 104-105, 108-109, 135, 149, 151-152, 154, 171, 185-188, 192-194
MSI devices, 85
Murray, A., 18, 148, 170-171

National Aeronautical and Space Administration (NASA), 2-3, 42, 146, 151, 153-155, 170; Apollo program, 149-150, 152
National Bureau of Standards, 4, 153
National Science Foundation (NSF), 64
National Semiconductor Corporation, 23-24, 40, 42-44, 48, 53-54, 57-58, 63, 82, 85-86, 88, 90, 92-93, 95-100, 103, 105, 107-108, 121, 135-136, 171, 186-187, 194-195
Naval Research, Office of, 152
Navy, Department of, 42-44, 153-154
Nesheim, John L., 165
Netherlands, The, 24, 167
New York Times, The, 92-93
Nippon Electric Corporation, 24, 64, 81, 93, 96, 136
Nitride Chemical Vapor Deposition, definition of, 182
Noncash bonuses, 193
Nonexperience-curve yield improvements, 34
Nonfinancial incentives, 190-191
Nontariff barriers, 161

North American Rockwell Corporation, 88
Northern Telecom Ltd., 136, 197
Novus Brand of calculators, 99
Noyce, Robert N., 26-27, 35, 47, 83, 132, 143, 185

Objectives, strategies, and tactics (OST) system, 54-55, 66
Occupational Safety and Health Organization (OSHA), 5
Off-the-shelf devices, 143
Offshore assembly plants, 161, 163
Oki Electric Corporation, 96
Opportunity: for innovation, 11, 13; market, 91, 147; for profit, 112
Options: product performance differentiating, 87; stock, 5, 67-69, 73, 164, 168-170, 172, 178, 189, 191-193, 196
Organizational: flexibility, 37, 45-46, 50-55, 69, 72; structure, 38
Original equipment manufacturers (OEMS), 89; sales, 99
Osborne and Associates, cited, 97-98
Overseas: income, 171-172; personnel, 169
Ownership: employee plans, 168; foreign, 145; stock, 174
Oxide masking, factor of, 16

Pacific Projects, Ltd., 132
Patents, 46, 62, 64, 68-69, 74, 103; applications, 60, 70; cost of filing, 63; holders of, 42, 107; licensing practices, 5, 145, 156-157; policies on, 38, 173; royalties, 156
Payback periods, 61, 135
Pennsylvania, University of, 152
Pension plans, 194
Performance: company, 6-7, 25, 102, 111, 117; competitive, 9; corporate, 67; economic, 165; feedback of, 111, 135-137; financial, 111, 135; improvement in, 177; international, 111, 118, 131-134, 142; past, 106; product differentiating, 87; real, 90;

Index

social, 111, 176; specifications, 147
Perkin-Elmer Company, 41
Personal service income, 166, 169
Personnel: key, 46; mobility, 17, 73, 103, 138; overseas, 169; procurement, 88; research, 170; skilled, 17, 38; training of, 88; top-notch, 168, 178
Philco-Ford Corporation, 17, 23, 47, 51-54, 85, 88, 133, 136
Philips Corporation, 23-24, 40, 42, 44, 81, 134, 145, 167
Photolithography-Projection Alignment, definition of, 183
Physics, 48; solid-state, 13
Planar process, 13, 16, 47, 180
Plants and equipment, 6
Plasma etching, definition of, 183
Plastic Encapsulation, definition of, 182
Policymakers, 1-2, 7, 141-142
Polysilicon, development of, 91, 183
Porter, Michael E., 103, 106
Pressure: competitive, 26, innovative, 167
Price competition, 77, 92, 100-104, 106, 137
Prices and pricing, 6, 77, 79; aggressive, 82, 94, 99, 100-101, 137; constraints, 26; cutting campaigns, 35, 84-85; discrimination in, 157; reductions in, 11, 102
Priorities, national, 170
Procurement: contracts, 2, 18, 74, 147-150; funding, 154; policies, 142, 145-151, 172-173
Producers: Japanese, 106; and surplus, 130-134
Product(s): consumer, 9, 25, 80, 98-101; cycles, 74; design, 78-79; development, 77, 102-103; digital integrated areas, 19, 88; electronic, 24, 99; life-cycle patterns, 77, 104; mixes, 2; performance differentiating options, 87; proprietary, 81, 92, 102, 106; standardization, 19, 94; strategies, 77-79, 102
Production, 66, 109; captive, 21, 80-81;
costs, 11, 30, 34, 48; engineers, 60; equipment, 34; materials, 6; military, 147; process, 13; silicon, 80; yield problems, 80
Profit and loss (P and L) criteria, 46, 53-54
Profits, 6, 61, 114, 166, 168, 176; above average, 33; company, 130; corporate, 112, 116; margins, 24, 79, 94; net, 117; opportunities, 112; potential, 26; sharing plans, 169
Programmable read-only memories (PROMs), 184
Promotion and public-relations campaigns, 87, 91
Proprietary: effects, 5, 171; designs, 21, 48, 78-79, 96, 135; integrated circuits, 84, 111; products, 81, 92, 102, 106
Public good, factor of, 65-66
Purdue University, 152

Quality: bonding, 78; management, 169

Radio Corporation of America (RCA), 23-25, 40-44, 51-52, 55, 63-65, 93, 96, 108, 133-134, 149, 151, 156, 193
Radio frequency transistor, 109
Rand Corporation, 157
Random access memories (RAM), 19, 34, 89-93, 111, 114, 119, 121, 124, 126, 129-130, 138, 184
Rates of return, target, 6
Raytheon Corporation, 23, 85, 96, 151
Read only memories (ROM), 19, 184
Recessions and depressions, effects of, 21
Regulations: antitrust, 170-171; governmental, 5; trade, 142, 158, 163
Reindel, Tom, 11
Reliability programs, 78, 153, 163
Rents, innovation, 131
Research: basic, 42, 46, 62-68, 74, 115, 155, 170; cost of, 64; efforts in, 150-153; in electronic, 155; flexibility, 154; funds, 5, 18; governmental, 4; personnel, 170; priorities, 177; re-

Research: cont'd.
  sources, 150; silicon-on-sapphire, 153; staff, 173; university, 155, 170-172
Research and development (R & D) programs, 1-2, 18, 45-48, 53, 59, 64-66, 111, 138, 144; expenditures, 5, 164-165; funding, 3-4, 142, 144, 151-155, 172-173, 177-178; skills and expertise, 6, 38; spending, 37, 56-62, 71-74, 170
Resistive metal deposition, definition of, 181
Resistor Capacitor Transistor Logic (RCTL), definition of, 84
Resistor Transistor Logic (RTL), 30, 84-85, 184
Resistors, 179-180
Resources, 45; base, 2; company, 38; engineering, 83; financial, 1, 6, 38; management, 79; research, 150; tangible and intangible, 6
Responsibilities, factor of, 53
Restrictions, legal, 73
Retrieval of information, 90-91, 122
Revenues: MOS, 114; royalty, 68
Reverse engineering, 102
Rewards, financial, 185, 189-190, 192
Rheem semiconductors, 171
Riley, Wallace, 92
Risk(s), 71-72; attitude toward, 6; return tradeoffs, 166; market, 38, 60-61, 74; taking of, 37-38, 46, 49-50, 56-62, 69, 72-73, 133, 152; technical, 38, 60-61, 74, 177
Rockwell International Corporation, 23, 25, 42-44, 95-96, 98-100, 107, 193
Rosenbloom, Richard S., 6
Royalties, 62-63, 107, 173; payments, 155-156; rate, 64; revenue, 68
Rules. See Regulations

Salaries: factor of, 67-69, 74, 196-198; rate of, 68; white-collar, 185-189
Sales, 6, 11, 13, 71-72; European, 131-133, 137; growth rate, 83, 88, 116; Japanese, 131-133; merchant, 21-22, 144; original equipment, 99; and output, 21; personnel, 66, 68, 80; ranking by, 22; worldwide, 1, 23-24
Sanders, Jerry, 48, 58, 196
San Francisco, California, 17
Sanyo Corporation, 99
Satellite communications, 143
Schlumberger Limited, 51, 145, 195
Schnee, Jerome, 148
Schottky: junction, 39, 182; low-power, 54; TTL, 85
Schuyler, Peter, 33
Schwartz, Nancy L., 56
Scientists and science, 13, 47
Search systems, 143. See also Research and Development
Second-echelon firms, 155
Second sourcing, 18, 21, 59, 78-82, 84-85, 88-89, 92, 94, 96, 106, 120, 137; agreements, 97; designs, 135; firms, 26; production, 48; strategy, 102
Secosem Company, 24
Security, reasons for, 4
Self-reinforcing feedback effects, 16
Semiconductor: firms, 1, 168, 174; industry, 163-167, 172, 177; memory products, 91; technology, 1-2
Semiconductor Electronic Memories Incorporated (SEMI), 91
Semiconductor Industry Association, 165
Sevin, president of Mostek Corporation, 198
SGS-ATES Company, 24
Shares and sharing, 114, 169
Sharp Company, 99
Shepherd, Mark, 47-48
Sherman Fairchild Foundation, 195
Shockley Transistor Company, 47
Sickman, Phillip, 84
Sideris, George, 91-92
Siemens, A.G., 24, 42-44, 81, 167, 196
Signetics Corporation, 23-24, 41-44, 53, 81, 84-85, 88, 93, 96, 108, 134, 145,

Index

149, 167, 169
Silicon: crystal, 24, 28, 30-31; devices, 147; oxide, 86; production, 80; purification, 24; transistors, 13, 16, 29, 148; valley, 17
Silicon-on-sapphire process, 41, 183; research, 153; technology, 93
Siliconix/American Microsystems, 40
Singapore, 163
Skills and skilled talents: engineering, 33, 174; labor, 5, 54, 174; personnel, 17, 38, 60, 171-172; research, 6
Social: benefits and welfare, 122; performance, 111, 176
Society, factor of, 6, 124
Solid state: electronics, 170; physics, 13
Sony Corporation, 64, 132
Sources: of capital, 1; multiple, 21, 40. *See also* Second sourcing
Southeast Asia, 34
Space: applications, 2; program, 23, 146, 151
Specialists and specialization, 19, 54, 66-67, 112
Specifications: military devices, 78, 143, 149; performance, 147
Speculation, factor of, 17
Spending habits, 37, 56-62, 71-74, 170. *See also* Expenditures, research and development
Spillovers, 155, 176; commercial, 143; real, 152
Sporck, Charles, 48, 58, 85
Sprague Electric Corporation, 23, 85, 198
Staffs: marketing, 37-38, 66-67; research, 173; sales, 68
Standard Microsystems, 40
Standard Telecommunication Labs, 41
Standards and standardization, 5, 19, 77, 88, 90-91, 94
Stanford University, 155
Status, cited, 22, 25, 94, 98, 150
Stereo Tape, magnetic, 90
Stock: capital, 165; holdings, 67, 189-191; options, 5, 67-69, 73, 164, 168-170, 172, 178, 189, 191-193, 196; ownership plans, 174
Storage capacities, 19, 34, 90-91
Strategy: competitive, 2, 5, 8, 50, 177; corporate, 81; diversity, 120-122, 137; feedback, 138; marketing, 82; mix, 176; product, 77-79, 102; second-sourcing, 102
Stress relief, annealing for, definition of, 181
Subsidies, 158
Substrate Diffused Collector Transmittor, definition of, 179
Suppliers and supply factors, 8, 25-26
Surfact barrier transistor, 47
Surplus: consumer, 122-130; producer, 130 134
Sylvania Corporation, 22-23, 40-41, 51-53, 82, 104, 107, 133, 137, 150-151
Sylvania Universal High-Level Logic (SUHL), 84-86

Taiwan, 163
Takeovers, 135-136
Tanaka, William H., 154, 161
Tariffs, foreign, 161
Tax Reform Act, The, 165-166
Taxation and taxes: capital gains, 5, 73-74, 164-169, 174, 190; credits, 170-171; exempt treatment, 169; income rate, 169; laws, 5, 169-170; policies, 2-5, 73, 119, 142, 164-165, 172, 178; property, 7, 9, 20, 152, 170
Technical: expertise, 38, 68; management, 46-50, risks, 38, 60-61, 74, 177; specialists, 66-67
Technology, 25-35, 50; advances in, 12; bipolar, 16-17, 90, 92, 94-95, 104, 115; changes in, 5, 11, 13, 16, 67, 167; commercial, 143; commitment to, 45, 72; digital integrated circuit, 13, 94; double polysilicon, 93; entrepreneurs in, 46-47, 74; fabrication devices, 143; foreign, 145; industrial, 171; memory, 94, 103; metal oxide semiconductor (MOS), 16-17, 70,

Technology, cont'd.
90-91, 95, 100, 104-105, 115, 136, 153; military, 143; new, 168-171; opportunities, 118; progress, 6; races, 83, 175; semiconductor, 1-2; silicon-on-sapphire, 93; VASIC, 164
Telecommunications: equipment, 156; industry, 13; systems, 24
Teledyne Corporation, 47
Telefunken Company, 24
Television industry, 18, 134, 189
Testing areas, factor of, 33-34
Texas Instruments Corporation (TI), 16-17, 21, 23-25, 40-44, 47, 51, 54-55, 57, 63-64, 66, 71, 80, 84-85, 87-89, 92-93, 96-97, 99-101, 104, 108, 132, 147-148, 152, 154, 186, 192, 194
Theft, employee, 171
Tilton, J., 4, 16, 23, 64, 70, 105, 132, 147, 151-152, 156, 173
Time path, ceiling-price on, 35
Tinkertoy Project, 4, 153
TMI variables, 71-72
Top-notch management and personnel, 38, 46-50, 70-73, 133, 168-169, 178
Toshiba Corporation, 24, 64, 96
Toys, children's, 18
Trade: associations, 6; barriers to, 111; foreign, 163; international rules, 142, 158, 163; literature, 8, 46; patterns of, 159-161; policies, 2-4, 158-164, 178; secrets, 63
Tradeoffs, policy of, 166
Training: engineers, 5; procurement personnel, 88; up-to-date, 171
Transistor(s): bipolar, 28-29; CMOS, 39; digital integrated circuit, 26-29; field-effect, 17; germanium-mesa, 13, 147; merged, 181; metal oxide semiconductor, 179; radio frequency, 109; silicon, 13, 16, 29, 148; surface barrier, 47; switching property of, 28
Transistor-Transistor Logic (TTL), 30, 52, 54, 96, 184; circuits, 119; families, 84-85; Schottky, 85; standard, 91; Sylvania's, 104

Transitron, 16, 23, 148, 150, 192
TRW Company, 193

Ultrasonic bonding, definition of, 181
Ultraviolet photolithography, 183
United Kingdom, 1
United States, 24
United Technologies Corporation (UTC), 81, 145, 198
Universities, 65, 68; government support of, 5; research in, 155, 170-172
Utterback, J., 18, 148, 170-171

Vacuum-tube industry, 22-23, 52
Veeco Corporation, 41
Venture capitalism and capitalists, 1, 17-18, 58, 142, 164, 166-168, 174
Vertical integration, 11, 21, 23, 25, 77, 80-81
Vertical MOS (VMOS), 181
Very large-scale integration program (VLST), 133
Video games, 19
VHSIC program: analysis of, 143-144, 174, 178; technology, 164

Wafer-fabrication: facility, 1, 18, 31-34, 132, 159; production area cleanliness, 78; slicing, 30
Walker, C.F., 168
Wall Street Journal, cited, 18, 168, 198
Warfare, electronic, 143
Watch industry, 18-19, 25, 28, 88, 98, 100, 101
Weapons-control systems, 143
Weather forecasting, 143
Weaver, Warren, 52
Webbink, Douglas W., 2, 12, 21, 23, 25, 156, 163
Weisburg, Leonard R., 143
Welfare, social, 122
Wells, Louis T., 104
West Germany, 136, 163. *See also* Germany
Western Digital Corporation, 97, 100, 105

Index

Western Electric Company, 4, 22, 24-25, 41, 63-65, 68, 80, 151-152, 156-157
Westinghouse Corporation, 22-23, 42-44, 51, 53, 85, 137, 149-150, 153, 156-157
White-collar: employees, 67; salaries, 185-189

Willig, Robert D., 129
Wire-bonding, 163
Wolff, Howard, 92-93
Worldwide: income, 170; market, 111, 147, 164; sales, 1, 23-24

Zenith Corporation, 134
Zilog Corporation, 96-98, 105

About the Authors

Robert W. Wilson is an economic consultant located in Weston, Massachusetts. He received the S.B. degree in physics from Massachusetts Institute of Technology and the Ph.D. in economics from Yale University. Dr. Wilson specializes in industrial organization and the economics of technological change and has conducted economic research for antitrust and regulatory proceedings. He has served with the Antitrust Division of the U.S. Department of Justice and as a senior research associate at Charles River Associates. He has written articles on the economics of technological change, government reports dealing with the price discrimination statutes and with regulation of the petroleum industry, and he coauthored *The Economics of Competition in the Telecommunications Industry*.

Peter K. Ashton, formerly a senior research associate at Charles River Associates, is currently employed as an associate at Putnam, Hayes and Bartlett, Inc. He received the master's degree in international economics and finance from Columbia University in 1978. Mr. Ashton's primary areas of interest include industrial organization, antitrust economics, the economics of technological change, and international trade and finance.

Thomas P. Egan is a senior research associate at Charles River Associates Incorporated in Boston. He received the B.S. degree in physics from Providence College in 1963, the M.B.A. from the University of San Francisco in 1975, and the Ph.D. in economics from the University of California at Davis in 1978. Dr. Egan held various marketing positions in the semiconductor industry before beginning his doctoral training. He has researched and coauthored studies in coal-market economics and the conversion of American industry to coal.